Mitsubishi

Lancer Evolution

The Rally Giants series:

Audi Quattro (Robson)
Austin Healey 100-6 & 3000 (Robson)
Fiat 131 Abarth (Robson)
Ford Escort MkI (Robson)
Ford Escort RS Cosworth & World Rally Car (Robson)
Ford Escort RS1800 (Robson)
Lancia Delta 4WD/Integrale (Robson)
Lancia Stratos (Robson)
Mini Cooper/Mini Cooper S (Robson)
Peugeot 205 T16 (Robson)
Saab 96 & V4 (Robson)
Subaru Impreza (Robson)
Toyota Celica GT4 (Robson)

www.veloce.co.uk

First published in October 2022 by Veloce Publishing Limited, Veloce House, Parkway Farm Business Park, Middle Farm Way, Poundbury, Dorchester DT1 3AR, England.
Tel +44 (0)1305 260068 / Fax 01305 250479 / e-mail info@veloce.co.uk / web www.veloce.co.uk. ISBN: 978-1-787117-12-9; UPC: 6-36847-01712-5.

RALLY GIANTS

Mitsubishi

Lancer Evolution

VELOCE

Brian Long

Contents

Foreword 5

Mitsubishi's WRC background 6

The fourth generation Lancer 13

The 1993 rally season 17

The 1994 rally season 21

The 1995 rally season 27

The 1996 rally season 35

The 1997 rally season 42

The 1998 rally season 52

The 1999 rally season 60

The 2000 rally season 70

The 2001 rally season 78

The 2002 rally season 92

The 2003 rally season 100

The 2004 rally season 102

The 2005 rally season 110

Mitsubishi's WRC record in the Lancer Evolution era 119

A 'LanEvo' swansong 121

Index 126

Foreword

Whilst researching my original Mitsubishi Lancer book, which covered the road cars as well as the competition machines, Rolf Eckrodt (who was in charge of the Japanese maker at the time), provided me with an incredible opportunity to sample some of the company's older rally cars to help put the progress made by the modern 'LanEvo' models into perspective. It was an unexpected surprise, although it certainly did the trick, highlighting how quickly the automobile had evolved within a relatively short space of time, and with remarkable clarity.

Dawdling along at the Okazaki test facility, the 1974 Safari Rally winner certainly had charm, but the little Lancer felt positively slow and fragile compared to the turbocharged model that came third in the 1982 1000 Lakes Rally. The first of the new generation Lancers made its debut as a road car in the autumn of 1991, ultimately becoming the chosen weapon in the works rally campaign soon after. The original 'LanEvo,' with its increased power and four-wheel drive, provided another jump in technology and pace, although even that struggled to prepare the author for the then-current version of the machine, with electronics helping to put the traction down and maintain stability even if you deliberately try to disturb the vehicle's balance. It was a real eye-opener, it has to be said …

Seeing as each car was a reflection of the era in which it was built to compete, this comparison test also revealed how much the face of world class rallying had changed over the same period. The pace had risen by a huge amount since the first World Rally Championship rounds of 1973, that much was obvious, and the materials used to build these machines had evolved in line with the Space Age, with lighter and stronger components being adopted with each passing season.

Here then, with the help of contemporary illustrations sourced from the factory, is the story of the works Lancer Evolutions in the World Rally Championship, along with the heroic folks who piloted them over everything from fast tarmac to broken paths, and in all weathers – a series of models that, from 1993 to 2005, went head-to-head with cars like the Lancia Delta, Ford Escort and Focus, Toyota Celica and Corolla, Peugeot 206 and 307, Citroën Xsara, and, of course, the legendary Subaru Impreza, in what is doubtless the most gruelling motorsport arena known to man.

Brian Long
Chiba City, Japan

Mitsubishi's WRC background

The first generation Lancer, introduced in early 1973, claimed a number of World Rally Championship victories. Familiar names in the Lancer's early rallying history included Andrew Cowan (who later became the Mitsubishi Ralliart boss), Joginder Singh and Kenjiro Shinozuka.

After a disappointing 1973 event (using the Galant), Joginder Singh and David Doig won the 1974 Safari Rally with the Colt Lancer, repeating the feat two years later as if to silence the critics. In fact, the tough little Mitsubishi took the first three places in 1976.

Andrew Cowan followed a string of good placings in the Safari (including a third in 1976) by winning the 1977 Bandama (Ivory Coast) Rally, again with the robust Lancer model. He was followed home by another Colt Lancer, driven by Mitsubishi team-mate, Joginder Singh. Although the latter event was not part of the WRC that year, it was incorporated into the Championship the following season, so it was definitely a world class rally. However, Mitsubishi was still not considered a major team in the 1970s.

The company suspended its rally programme in light of the second major oil crisis to hit during the decade, although 1980 was a key year, as European headquarters were established in Austria to co-ordinate the company's international competition activities. Following this, Mitsubishi duly launched a serious attack on the WRC crown with a two-litre turbocharged version of the second generation Lancer.

The turbocharged Lancer

The new Lancer appeared on the 1980 Motogard Rally of New Zealand (finishing tenth in the hands of Morrie Chandler), but it wasn't until the 1981 Acropolis Rally – held in June – that the works-entered, 280bhp turbocharged cars made their debut.

Anders Kulläng, who had a long career with Opel, was one of the official team drivers, as was Andrew Cowan, the Mitsubishi stalwart who'd just completed a stint with Mercedes-Benz after the Japanese company reduced its

Action from the 1974 Safari Rally.

WRC efforts towards the end of the 1970s. Sadly, both cars (homologated as Group 4 machines) retired early on, one with fuel problems, the other with a cooked alternator.

Undeterred, the team travelled to Finland for the 1000 Lakes, this time with three cars. There were problems once again, but, with the cooler temperatures, all three finished, with the top one (driven by Antero Laine) taking tenth place. Ultimately, the best result of the year came in the last event of the season – the 1981 RAC Rally – when Kulläng claimed ninth.

A revision of the homologation groupings (from Group 1 to 4, to the new Group A, B and N) saw the face – and pace – of rallying change forever. Introduced in 1982, the era of the Group B supercars had begun, but with Mitsubishi having such a small operation dedicated to the rallying scene, the company was not able to take advantage of the new rules as quickly as it would have liked.

However, Anders Kulläng was joined by 'Flying Finn' Pentti Airikkala as a Team Ralliart driver for the 1982 1000

Lakes Rally, when Mitsubishi returned to the arena; Airikkala finished a gallant third. A few weeks later, Kulläng claimed seventh in the San Remo event, and, on the Bandama, a lowly Group 2 Lancer (driven by Eugene Salim) was one of only six cars to finish.

The RAC was a disappointment (Airikkala crashed out, and Kulläng had clutch problems), but the turbocharged machine was starting to come good. It was too late, however, as all top entrants had to comply with the new Group A/B/N regulations for 1983 (older cars had been allowed to compete in 1982, and, oddly, privateers could continue using them for a little while longer). Salim took sixth on the Bandama (Ivory Coast) Rally again, but that was to be the best result of the season.

Rally raids

The Mitsubishi Pajero has an enviable reputation in long distance marathons, the most famous of which being the gruelling Paris-Dakar Rally, one of the toughest challenges on the face of the Earth. The Pajero made its debut on this event in 1983 (shortly after it was introduced), and duly

A Lancer on the 1983 Cyprus Rally, part of the European Championship for that year.

won its Class. It took Class honours again the following year, before taking overall victory in 1985. This was to be the first of many wins for the Pajero in the desert.

The Starion

Eventually, in mid-1984, a four-wheel drive version of the Starion made an appearance on the Mille Pistes in France, and duly won the Prototype category. The Starion was later developed, albeit with rear-wheel drive, by Ralliart (at that time centred in Maldon in Britain, but later moved north to Rugby) into Mitsubishi's new WRC contender for 1986. The Starion was a Group A machine, although appearances at the 1000 Lakes, Ivory Coast and RAC Rallies gave little in the way of success. Meanwhile, Salim took fifth on the 1985 Ivory Coast Rally in his locally registered Lancer Turbo, and Martial Yace picked up a top ten spot the following year.

Although it achieved some reasonable results at international and national levels, the Starion was less than successful in the ultra competitive WRC events. Fifth place on the Rally of New Zealand (in the hands of privateers, David and Kate Officer) was the highlight of 1987 until the Ivory Coast event, when the big coupés finished fourth and seventh. The Officers took fourth in NZ in the following year, while Patrick Tauziac claimed third in the Ivory Coast (he also took second place with the same model in 1989). However, the days of the Starion Turbo were numbered.

The Mitsubishi Pajero of Patrick Zanirolli on its way to a fine victory on the 7th Paris-Dakar Rally (1985). Class honours had been collected in the two years prior to this, but this was the first of many overall wins.

Mitsubishi entered a works team on the 1984 RAC Rally, but failed to get on the leader board. The 4WD Starion was an interesting experiment, however.

Below: A more conventional Starion on the 1986 1000 Lakes Rally. Antero Laine actually won a couple of rounds of the Finnish Championship with the Starion that year.

Shinozuka on his way to victory on the 1991 Ivory Coast Rally. Early Galant VR-4s used a viscous-coupled centre differential with a 50:50 split. By the end of the car's WRC reign, it employed electronic control for the rear and centre differentials, with a 30:70 split in favour of the back wheels.

The Galant era

The Starion Turbo era was brought to an end with the announcement of the four-wheel drive, four-wheel steer Galant VR-4. Making its debut on the American Olympus Rally (which started on 23 June 1988), the Group N machine showed great promise, and finished tenth overall.

Kenjiro Shinozuka – ultimately the Asia-Pacific Champion of 1988 – took a Group A car to New Zealand, only to have gearbox problems, but Ari Vatanen gave the

290bhp Galant a good run in the RAC until the engine packed up due to an electrical fault.

Vatanen, the ace all-rounder, was very quick in the 1989 Monte Carlo Rally, but finished a poor (at least by his high standards) 87th. Although Vatanen retired on the Acropolis, Jimmy McRae (Colin McRae's father) came home fourth in Greece, with Shinozuka following close behind in seventh; the Galant VR-4 also took Group N honours.

Good results were achieved again in New Zealand, and then, at last, Mikael Ericsson and Claes Billstam (who started the season with Lancia) won the 1989 1000 Lakes Rally. Later, decent results were attained in Australia, when Pentti Airikkala and Ronan McNamee took a surprise victory on the RAC Rally to round off the season nicely (Vatanen came home fifth in a similar Mitsubishi Ralliart Europe car).

I remember watching the RAC that year, thinking that the Japanese had arrived in a big way. Nissan (Datsun) had started the ball rolling and now it seemed unstoppable, as both Toyota and Mazda had extremely competitive cars, and Mitsubishi had also proved it was capable of winning at world class level. The traditional dominant marques in the WRC arena could no longer rest on their laurels. As with the arrival of the Scandinavian drivers a couple of decades earlier, ardent followers of the rally scene were witnessing the dawn of a new era.

The Galant VR-4 was developed further over the 1990 season, gaining a viscous-coupled centre differential towards the end of the year. A poor start in Monte Carlo (both cars retired) was temporarily forgotten in Portugal, with the Galants leading one-two. However, both works machines failed to finish. It seemed as if a fifth on the Safari for Kenjiro Shinozuka broke the run of bad luck, but, sadly, there were more retirements in Greece.

Ross Dunkerton took fourth place in New Zealand, the rally on which a new name joined the Mitsubishi camp – Tommi Mäkinen. The Finn claimed Group N honours in his Galant VR-4, coming home sixth overall in his first WRC event.

During the 1000 Lakes Rally, it was announced that Ari Vatanen was leaving Mitsubishi, which was perhaps more than a touch ironic, as he finished the event in second place. He was followed home by Kenneth Eriksson (third), Lasse Lampi (seventh), and Tommi Mäkinen (11th overall, and first in Group N once again).

Mäkinen continued his winning ways in Australia (Eriksson retired with transmission problems in the top Group A car), and Patrick Tauziac (a true Mitsubishi stalwart, co-driven by Claude Papin) won the 1990 Ivory Coast Rally.

At the last event of the season, the RAC Rally in Great Britain, Eriksson came second, while Vatanen, who had achieved great speed during the early stages, retired on his last Mitsubishi drive (he later signed up with Subaru). Relatively consistent results ultimately gave Mitsubishi third place in the World Championship that year.

For 1991, the Galant adopted bigger wheels and tyres, which enabled the engineers to fit larger brakes. The driver line-up included Kenneth Eriksson (who remained with the Japanese team from the previous season) and Timo Salonen. Salonen claimed eighth on the Monte Carlo, but Eriksson retired after running out of fuel. However, the Swede won his home event, thus giving the VR-4 its third WRC victory.

The 1991 Acropolis Rally saw the birth of the Galant VR-4 Evolution, developing over 300bhp from its two-litre engine. The larger intercooler and louvred power bulge on the bonnet were identifying features, but Salonen earned a DNF for his record, while Eriksson's chances were ruined by a seven minute penalty. Nonetheless, the car had shown a great deal of promise.

On the 1000 Lakes, Tommi Mäkinen got his first works drive, although ironically it was for Mazda. Salonen claimed third on the road, but was disqualified, thus handing the position to his Swedish team-mate. Mäkinen, incidentally, came home in fifth, and switched allegiance to Nissan for 1992.

The Mitsubishi pair came second and fifth in Australia (with Ross Dunkerton seventh), and there was a win for Shinozuka on the 1991 Ivory Coast Rally. Tauziac came second, and, with good placings on the RAC as well, it was enough to give the Japanese company third in the World Championship again.

Eriksson and Salonen continued for the 1992 season, although a fifth place in Portugal was the best they could muster after the first six events. Ross Dunkerton came to the rescue with a podium finish in New Zealand, and took fifth in Australia. As some form of consolation, Shinozuka guided the Galant VR-4 to another win on the Ivory Coast Rally, but there was no doubt that 1992 had been a disappointing season.

Although the Galant had performed admirably, there was no escaping the fact that it was a big, heavy car compared to those fielded by the competition. To be in with a chance of winning the World Rally Championship, it was obvious that Mitsubishi needed a lighter, faster and more nimble machine. Enter the Lancer Evolution series ...

The changing face of rallying

Rallies have been around in one form or another since the dawn of motoring itself. Classics like the Monte Carlo Rally established a familiar format that included a series of timed stages in an event lasting several days. As the rallies gradually got tougher, makers saw the marketing value in having cars reliable and fast enough to compete.

Fast forward to 1973, when the FIA World Rally Championship was introduced, and a win in an international rally was ultimately seen as the ideal promotional tool for manufacturers of sporting machinery. Back then, most of the vehicles were fairly standard – the sort of cars you could buy in the showroom – although the 1982 season saw the birth of the Group B era, with 200-off monsters dominating the scene; Group A was for 5000-off cars with substantial modifications, and Group N for similar models with hardly any changes to the production specification.

However, the Group B cars were banned at the end of 1986 due to a series of accidents that highlighted the dangers of running outrageously fast machinery in a rally environment. Group A and Group N cars became the norm, with the homologation requirement eventually cut to just 2500 vehicles. Nonetheless, to be competitive in rallying still called for engines developing over 250bhp and 4WD systems to put the power down effectively. To encourage more makers to take up the challenge, the FIA introduced a World Rally Car category (to run alongside Group A and Group N) in time for the 1997 season, with only 20 identical two-litre turbocharged machines required for homologation, and things became even more specialised after the requirement was reduced to just ten cars ...

The fourth generation Lancer

The sporting Lancer had always had a strong following amongst enthusiasts, coming of age via a turbocharged variant based on the second generation model in the autumn of 1981. As well as the Mitsubishi Colt brand fielded in the UK, numerous Lancers found their way into Dodge and Plymouth dealerships in the States (thanks to a tie-up with Chrysler) and, later, others were branded as Eagles, Hyundais and Protons.

The fourth generation Lancer made its debut at the 1991 Tokyo Show, sold only in four-door saloon guise by this time. As before, a lot of high-tensile and galvanized steel was used in the monocoque body construction, giving it strength and longevity. With a wheelbase of 98.4in (2500mm), the overall length was 168.3in (4275mm), the width 66.5in (1690mm), and the height 54.5in (1385mm); the track measured 57.1in at the front and 57.5in at the rear (1450 and 1460mm respectively). Therefore, the new car was longer, wider and lower than its predecessor. The Cd was listed at just 0.31.

There were six new engines, including a V6, although it was the fuel-injected four used in the RS and GSR grades that interests us – a 16v 1.8-litre (81mm x 89mm) unit with twin cams, a turbocharger, and an intercooler. Rated at 195bhp, it was obvious that Mitsubishi had its sights on the WRC with this one, as both five-speed models fitted with this engine came with four-wheel drive as standard.

As for the suspension, MacPherson struts were employed up front, along with an anti-roll bar. The front suspension was mounted on a reinforced subframe, which was then attached to the body via elastomer bushes. This arrangement increased the rigidity of the mounting points

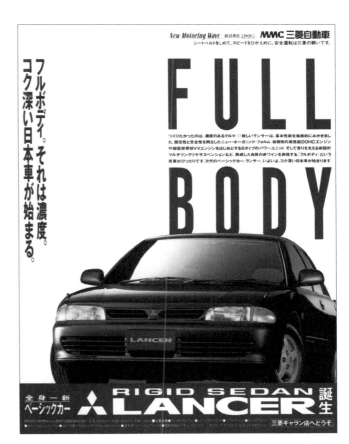

Advertising proclaiming the arrival of the fourth generation Lancer. The vehicle's Chief Engineer was Kazuyoshi 'Ikko' Kobayashi, and the Chief Designer was Shigehisa Ishii.

whilst insulating passengers from shocks and jolts.

At the back, there was a newly-developed multi-link setup with trailing arms. The upper and lower links minimised camber changes for enhanced cornering power (and traction on 4WD versions), and a toe-control link allowed a certain amount of passive rear wheel steering. Combined with a wider track and longer wheelbase, there was a marked increase in straight-line stability.

Steering was via rack and pinion, and the brakes were larger and more responsive than before; ABS was an option for all cars. The GSR came with 14-inch alloys, aerodynamic body parts, and Recaro seats.

Announced a few weeks ahead of the Subaru Impreza, Mitsubishi launched the Type CD9A Lancer GSR Evolution on 7 September 1992, with sales beginning in the following month. This would be the company's weapon of choice in the rally arena, with a good deal more grunt emanating from its two-litre lump. In order to ensure that the minimum 2500 units could be built for homologation, it had been decided to market the GSR alongside the stripped-out RS, as the management feared demand would be slow for a pure motorsport machine. While this may have been true, as the idea of making the Evo available for export markets was not entertained, the fears were ultimately unfounded, and the first batch (a mix of GSR and RS grades) was sold out in just three days, leading to a second batch of 2500 cars having to be constructed.

The body was light, compact, and strong. Reinforced in strategic areas, it had 20 per cent better torsional rigidity than the standard models, and the front/rear balance was also better through the use of an aluminium bonnet, which featured air intake/outlet ducts (the louvres in the bonnet were for hot air extraction). There was a large aperture in the front mask for improved cooling, and a big rear spoiler (incorporating a high mount rear brakelight) for increased downforce. Overall the Evolution was 1.38in (35mm) longer, and a fraction wider (at 66.7in, or 1695mm). Otherwise, it was the same dimensions as the production 4WD cars.

The engine was basically a modified Galant VR-4 unit – the Cyclone 2000 dohc 16v Intercooler Turbo (code 4G63). Bore and stroke measurements of 85 x 88mm gave a cubic capacity of 1997cc; combined with a large capacity intercooler, a new lightweight crankshaft, new pistons and con-rods, new injectors, revised port shapes in the head, sodium-filled valves, and a hike in the compression ratio (raised from 7.8 to 8.5:1), it delivered 250bhp at 6000rpm, and 227lbft of torque at 3000rpm.

Reduced internal friction gave better throttle response, especially at high revs. There was a big bore exhaust system,

Above and opposite: Cover and pages from the first Evolution catalogue.

used to drive a TD05H-16G-7 turbo (the '7' representing the diameter of the turbine nozzle), which featured an Inconel turbine (a mix of nickel, chrome and iron, with a trace of carbon); twin pipes exited from the rear. Because of the harsh environment in which the car would be used, an oil cooler was specified as standard.

Naturally, in view of its sole purpose, an automatic transmission was not even an option. Instead, the Evolution came with a close-ratio gearbox (with 2.571 on first, 1.600 on second, 1.160 on third, 0.862 on fourth, and 0.617 on fifth); the standard final-drive ratio was 5.443:1. An uprated clutch was employed, along with a double-cone synchro on second. A viscous-coupled centre differential was used on the full-time 4WD system, with an lsd at the rear – both items were borrowed from the VR-4.

The suspension was based on that of the standard

GSR but uprated, and given pillow-ball bushings instead of rubber ones at the back. Anti-roll bars were employed at both ends. However, the ride was deliberately not too hard for normal road use.

Ventilated discs were specified up front (with two-pot calipers), while solid discs were used at the back; four-wheel ABS came as standard. The GSR had 15-inch alloy wheels shod with 195/55 VR15 Michelin XGT tyres; fairly narrow, admittedly, but there was a distinct problem clearing the wheelarches with fatter rubber.

Maker options included a front limited-slip differential, an electric sunroof and Cibie foglights, while dealers were able to offer heavy-duty mudguards, a front strut brace, auxiliary gauges and a centre console-mounted kneepad.

At 2574lb (1170kg), the Lancer RS Evolution was about 155lb (70kg) lighter than GSR, as it was missing the ABS, air-conditioning, most of the electrical goodies, the rear wash/wipe, and various trim pieces. The Recaro seats were replaced by basic items, a mechanical rear lsd was employed in place of the viscous one fitted to the GSR, and it came with steel wheels instead of alloys.

Available in white or silver shades only, maker options included alloy wheels, a front lsd, a rear wash/wipe and Cibie foglight package. Dealer options included high-back Recaro seats (different to those fitted to the GSR Evolution), and the same items as offered on the GSR grade. This new, clearly defined role for the Lancer allowed the Galant to go further upmarket.

Another illustration from the first Lancer Evolution brochure.

The 1993 rally season

As mentioned earlier, the Galant was a little too big to be a serious WRC contender. The 1992 car was some 132lb (60kg) lighter than its predecessor, but was still too heavy to compete with the leading Lancias and Toyotas in faster events. However, there was no doubt that the Galant VR-4 engine and transmission was a superb combination, and, suitably adapted for use in a smaller, lighter shell, it held great promise.

Fortunately, Mitsubishi had the car it needed already in production – the Lancer. Equipped with the Galant's running gear, it had the potential to be a world beater. Ralliart Europe did some testing with a converted standard car, whilst the Japanese development team in Okazaki worked with a pre-production prototype. It had a different rear suspension compared to that of the Galant, and the rear-wheel steering was dropped in order to further save weight, although the intercooler was bigger than the VR-4's. With a shorter wheelbase, the new car was much lighter (it would ultimately be lighter than all of its WRC contemporaries), with far better weight distribution. It was homologated on 1 January 1993 (number A5469).

Kenneth Eriksson's contract was renewed, but Timo Salonen was replaced by Armin Schwarz for the 1993 season. Schwarz, born in Germany in 1963, had made his name with TTE in the early 1990s. The driving styles of these two were quite different: Eriksson was better on the rougher roads and Schwarz was a tarmac specialist. So with a new car to develop at the same time, quick results could not be expected.

As a matter of interest, Eriksson had also come from Toyota, joining Mitsubishi in 1990 during the Galant era.

Born in Sweden in 1956, Eriksson had been rallying since 1977, becoming full-time in 1986, the year in which he was declared World Champion (Group A).

1993 rally review

The Lancer Evolution made its Group A debut on the 1993 Monte Carlo Rally. Schwarz was the hard charger on this event, but transmission problems would ultimately cost him. However, there had been little time to test the Lancer, and the fact that Eriksson (who'd had his share of trouble) finished fourth, could only be regarded as a good start.

Due to budget constraints, Mitsubishi Ralliart entered only seven of the 13 rounds in 1993 (five, including the Monte, the Rallye de Portugal, the Acropolis, the 1000 Lakes and the RAC, were the responsibility of Ralliart Europe). In Sweden, the Mitsubishi effort was left to Group N Galants nominated as team cars, but in Portugal, the works Lancers were present once more. Unfortunately, Schwarz had a disastrous rally, with a wheel giving way on the first day, and an accident that put him out of the event. At least Eriksson's fifth place and the top ten finishes for the Galants in Sweden kept the Japanese company in the title chase.

A single car was sent from Japan for the Safari (entered by Mitsubishi Oils), but it retired with engine trouble after completing only 30 of the 79 stages. There was better news in Greece, though, with the first podium finish for the Lancer Evolution, running with a slightly wider track on the event. However, with Eriksson retiring on the first day with a broken sump, Toyota and Ford were pulling away in the Championship race.

1993 Monte Carlo Rally (Eriksson).

1993 Rallye de Portugal (Schwarz).

1993 Acropolis Rally.

1993 Safari Rally (Shinozuka).

The rival teams – Lancia

Rally fans were always treated to something special from Lancia throughout the seventies, eighties, and early nineties. The pretty little Fulvia gave way to the Stratos, and the Lancia Rally 037 kept the maker's momentum going into the Group B era. Then came the 4WD Delta S4 and HF Integrale, which gave the Italian manufacturer its tenth WRC title in 1992 – a tally that has yet to be beaten. However, while Lancia was still the dominant force that year, hardly dropping a point, the Martini-sponsored works team was disbanded soon after, and Toyota duly rose to the top of the pile.

1993 RALLY RECORD

No	Driver/co-driver	Position	Reg. No. (Group)
Monte Carlo (21-27 January)			
8	Kenneth Eriksson/Staffan Parmander	4th	K4 MRE (Gp A)
4	Armin Schwarz/Nicky Grist	6th	K5 MRE (Gp A)
Portugal (3-6 March)			
3	Kenneth Eriksson/Staffan Parmander	5th	K4 MRE (Gp A)
7	Armin Schwarz/Nicky Grist	dnf	K5 MRE (Gp A)
Safari (8-12 April)			
4	Kenjiro Shinozuka/Pentti Kuukkala	dnf	KNY70-HA7863 (Gp A)
Acropolis (30 May-1 June)			
9	Armin Schwarz/Nicky Grist	3rd	K7 MRE (Gp A)
4	Kenneth Eriksson/Staffan Parmander	dnf	K6 MRE (Gp A)
New Zealand (5-8 August)			
9	Ross Dunkerton/Fred Gocentas	dnf	SB 682 (Gp A)
1000 Lakes (27-29 August)			
3	Kenneth Eriksson/Staffan Parmander	5th	K5 MRE (Gp A)
6	Armin Schwarz/Nicky Grist	9th	K6 MRE (Gp A)
Australia (18-21 September)			
8	Ross Dunkerton/Fred Gocentas	4th	8NY 854 (Gp A)
RAC (21-24 November)			
4	Kenneth Eriksson/Staffan Parmander	2nd	K5 MRE (Gp A)
7	Armin Schwarz/Peter Thul	8th	K4 MRE (Gp A)

In New Zealand, Yoshio Fujimoto came home tenth overall in his Group N 'LanEvo,' but it was Dunkerton's Group A machine that was the nominated car – it failed to finish. Colin McRae chalked up his first WRC victory, putting Subaru ahead of Mitsubishi for the first time during the season.

As the circus moved to Finland, hopes were high. However, despite increased power, both cars struggled to keep up with the scorching pace – Eriksson won just one of the 35 stages, and Schwarz was never able to trouble the top runners. They finished fifth and ninth, but now Lancia had overtaken the Japanese team as well.

Reigning Asia-Pacific Champion Ross Dunkerton came fourth in his native Australia, although he was a long way down on the winning Celica. The RAC Rally, on the other side of the world, provided the Lancer with its last chance for a win. Sadly, the forests of Britain were not kind to Mitsubishi, but a second and eighth place was extremely encouraging nonetheless, especially in view of the strong competition in the World Rally Championship at that time.

Ultimately, Mitsubishi finished 1993 in fifth position, with Eriksson as the top driver, claiming seventh. However, Carlos Sainz was later unlucky enough to be excluded from San Remo, which had the effect of promoting the Japanese manufacturer to fourth!

1993 RAC Rally (Eriksson).

Kenjiro Shinozuka.

The 1994 rally season

The Lancer Evolution II (Type CE9A) was announced at the end of December 1993, with GSR and RS grades sold alongside each other from mid-January. Incredibly, all 5000 Evo IIs built were sold by the time April came around, and although never officially exported, a few did find their way abroad.

The Evo II featured a deeper front air dam, and a base section was added to the rear spoiler. It had a wider track at both ends, and also a slightly longer wheelbase, as the front wheel centres were moved forwards a touch. Most of the 22lb (10kg) weight increase could be assigned to the 30

Kenneth Eriksson.

Armin Schwarz.

per cent improvement in torsional rigidity brought about by reinforcements in the body.

The engine was the same 4G63 unit with an 8.5:1 compression ratio. The turbo was the same as the original Evo, but an increase in boost pressure, a better, freer-flowing exhaust system, and more lift in the valves gave 10bhp more. While torque output remained unchanged, power was now quoted as 260bhp at 6000rpm. At the same time, an air-cooled oil cooler was adopted.

The transmission had changes to first and second on the close-ratio gearbox (now 2.750 on first, and 1.684 on second). All the other ratios and final-drive remained the same, although third and fourth gained double-cone synchronisers; the clutch plate material was also of a higher quality. At the back, the rear lsd was now a 1.5-way mechanical unit on both RS and GSR grades.

The suspension geometry was subtly revised, with stronger mounting points. The lower arm at the front was now forged for added strength, the front anti-roll bar diameter was reduced from 0.91 to 0.63in (23 to 16mm), while the rear spring rate was increased.

Fatter 205/60 HR15 tyres mounted on OZ five-spoke aluminium alloys were now standard fare on the GSR, while the same rubber came on steel wheels for the RS. Combined with the suspension changes, this helped to improve cornering. There were also new brake pads, and the steering ratio was changed (from 16 to 14.8 to make it quicker), along with the power steering pump.

1994 Safari Rally (Shinozuka).

1994 Acropolis Rally (Schwarz).

1994 Rally of New Zealand (Schwarz).

1994 RAC Rally (Holderied).

On the WRC front, post-season analysis had established a problem with the Lancer's centre differential. Whilst the car was airborne, the hydraulic mechanisms were shifting torque between axles unnecessarily, thus reducing traction when it landed again. Subtle adjustments made the vehicle far more competitive, although the Evolution II was expected to take over the challenge in mid-1994.

The Evo II's debut duly came on the Acropolis. A number of aerodynamic changes made the vehicle more stable at high speed; with the addition of new shock absorbers, these were the main differences between the original Evolution model and its successor; even the homologation number was the same.

The FIA introduced a rotation system for the 1994 WRC, taking in ten rounds. Makers had to enter nine of these to qualify for the Manufacturers' crown, although halfway through the season, Mitsubishi revised its programme, meaning it could not be classified in the title chase.

Eriksson and Schwarz continued as Mitsubishi's works drivers, but Schwarz lost Nicky Grist (he joined Juha Kankkunen at Toyota towards the end of the last season) – Klaus Wicha, who had been with Schwarz in his early Toyota days, took Grist's place in the co-driver's seat.

1994 RALLY RECORD

No	Driver/co-driver	Position	Reg No (Group)
Monte Carlo (22-27 January)			
9	Kenneth Eriksson/Staffan Parmander	5th	L4 MRE (Gp A)
5	Armin Schwarz/Klaus Wicha	7th	L5 MRE (Gp A)
Portugal (1-4 March)			
25	Isolde Holderied/Tina Thorner	11th	GG-RL117 (Gp N)
19	Jorge Recalde/Martin Christie	dnf	GG-RC50 (Gp N)
Safari (31 March-3 April)			
4	Kenjiro Shinozuka/Pentti Kuukkala	2nd	KNY70-HA7863 (Gp A)
Tour de Corse (5-7 May)			
30	Isolde Holderied/Tina Thorner	16th	GG-R379 (Gp N)
Acropolis (29-31 May)			
8	Armin Schwarz/Klaus Wicha	2nd	K7 MRE (Gp A)
4	Kenneth Eriksson/Staffan Parmander	dnf	K6 MRE (Gp A)
Argentina (30 June-2 July)			
11	Jorge Recalde/Martin Christie	5th	GG-RC50 (Gp N)
14	Isolde Holderied/Tina Thorner	8th	GG-R379 (Gp N)
New Zealand (29-31 July)			
4	Armin Schwarz/Klaus Wicha	3rd	L6 MRE (Gp A)
8	Kenneth Eriksson/Staffan Parmander	4th	L5 MRE (Gp A)
17	Jorge Recalde/Martin Christie	dnf	GG-RC50 (Gp N)
1000 Lakes (26-28 August)			
15	Jarmo Kytolehto/Arto Kapanen	8th	X2177 (Gp N)
36	Olli Harkki/Antti Virjula	10th	H4233 (Gp N)
33	Isolde Holderied/Tina Thorner	29th	GG-RL117 (Gp N)
16	Jorge Recalde/Martin Christie	dnf	GG-RC50 (Gp N)
San Remo (9-12 October)			
25	Isolde Holderied/Tina Thorner	15th	GG-R379 (Gp N)
2	Tommi Mäkinen/Seppo Harjanne	dnf	K4 MRE (Gp A)
7	Armin Schwarz/Klaus Wicha	dnf	K5 MRE (Gp A)
RAC (20-23 November)			
30	Isolde Holderied/Tina Thorner	16th	GG-R379 (Gp N)

The pace of the Mitsubishis took many by surprise when the season opened in Monte Carlo. Armin Schwarz led the event until unruly spectators caused him to go off the road. Although both Mitsubishi drivers were devastatingly quick, claiming 11 of the 22 stages between them, they could not keep it up, and it was Ford that claimed the victor's laurels. However, Isolde Holderied and Tina Thorner took the Ladies' title in their Group N 'LanEvo.'

Group N cars were entered from Germany for Portugal, the Holderied/Thorner pairing again taking Ladies' honours, and finishing 11th overall – the top Mitsubishi. On the Safari, Kenjiro Shinozuka was again sent to compete in the Mitsubishi Oils-entered, Japanese-prepared Lancer. He put up a gallant effort, finishing second behind the hard-charging Toyota of Ian Duncan.

The Tour de Corse was covered by the Germans again, but the Acropolis saw the return of the MRE cars – brand new Evolution II models. Both Eriksson and Schwarz had minor troubles before Eriksson went out with a broken rear suspension. However, Schwarz gave the Evo II an impressive debut, taking second place at the end of the event.

It was up to the German team to once again uphold Mitsubishi honour in Argentina, and they did a sterling job, taking fifth and eighth. The big guns returned in New Zealand, ably supported by Recalde (who went out after an accident) and a number of privateers. A lack of power and the wrong tyres offset the improved traction of the works cars, preventing them from winning, although third and fourth was hardly shabby.

The 1000 Lakes saw a mixed bag of Lancers, entered alongside Galants by Ralliart Finland and Mitsubishi Germany. The end result was a win in Group N, and the Ladies' Drivers' title for Isolde Holderied (she only had to start in San Remo).

Ralliart Europe fielded cars in San Remo – only the organisation's fourth, full-blown works assault on the WRC during the season. It was a new event for MRE, with a new driver, too. The latest recruit was Tommi Mäkinen, who had done so well in the Group N Galants before moving into Group A. Born in Finland in 1964, he came to have a great influence on Mitsubishi's fortunes.

Anyway, San Remo was ultimately not a success, as Mäkinen's suspension collapsed, and Schwarz (the other works driver on the event), having claimed two stage wins, was put out after his car caught fire. At least Holderied won the Ladies' Class to take her title in style.

Holderied was in action again on the RAC, when she was hoping to claim to Group N Championship to add to her other trophy. Unfortunately, her main rival finished ahead of her, thus putting an end to her dream of a double. Interestingly, while MRE stayed away, Proton made its European rallying debut at this event. Subaru won the RAC Rally, but Toyota won the World Rally Championship.

The 1995 rally season

The Evolution III was launched just in time for the new World Rally Championship season. Carrying the same CE9A codes as its predecessor, the Lancer GSR Evolution III was launched alongside its RS stablemate in January 1995, and went on sale from 10 February.

New pistons gave an increase in compression ratio

進化は、とまらない。

WRC（世界ラリー選手権）で鍛えたエボリューションに、
さらにパワーアップしたⅢ、完成。

ランサーが、再びWRCにエントリーして、3年。
「走る・曲がる・止まる」というクルマ本来の走りの機能を、妥協を許さないステージで極限まで鍛えあげていった
そしていま、その数多くの挑戦から獲得したさまざまなノウハウをフィードバック。
三菱の新たな情熱を傾けた一台が誕生した
ランサーGSRエボリューションⅢ。
最高出力270PSを達成した卓越したパワー そして、風の流れを巧みに利用する革新的な空力ボディ。
スポーツセダンとして純粋な進化を遂げ、さらに最強の走りへと到達した
その走りの向こうには、輝かしい栄光が待っている

A page from the Evolution III catalogue.

(from 8.5 to 9.0:1) and brought an extra 10bhp, taking maximum power to 270bhp at 6250rpm (torque output was again unchanged). The turbocharger (now type TD05H-16G6-7) and exhaust system also received attention to give better response, while two water spray jets were employed on the intercooler instead of just the one.

Gearbox ratios were the same, but the final-drive was now fractionally higher at 5.358:1. All ratios were quite different to those of the standard GSR, incidentally. Wheels and tyres were carried over from the Evolution II, but there were more changes for the body. Indeed, all of the aerodynamic appendages were modified slightly.

There was a new front air dam with cooling ducts for the brakes and transfer box, and a taller rear spoiler incorporating a high mount brakelight in the base. Thanks to these modifications, lift was said to be -0.01.

While dimensions remained the same, there was another 10kg (22lb) weight increase, which makes sense, as a friend that owned an Evo II from new had stated to the author that the body was still not quite strong enough

1995 Monte Carlo Rally (Mäkinen).

for competition work, despite the improvements made over the original. Now, even the most demanding of buyers were left with little to complain about.

Changes were afoot on the WRC front. For 1995, manufacturers were forced to enter each of the eight WRC events. Ralliart Europe simply didn't have the finance to do this on its own, so in those rounds MRE didn't cover, Ralliart Germany's Group N car was nominated as the works entry. There was also a new scoring system, in which

1995 Swedish Rally (Eriksson).

1995 Rally of New Zealand (Eriksson).

1995 Catalunya Rally (Aghini).

Tommi Mäkinen.

Andrea Aghini.

the points from two cars were put forward for each event. Consistency and teamwork were more important than ever.

On the technical front, new, smaller bore restrictors limited airflow going into the turbocharger, thus levelling the playing field – all the major teams were getting around 300bhp from their two-litre engines. Mitsubishi introduced electronically-controlled front and centre differentials for 1995, representing another challenge for MRE's Chief Engineer, Roland Lloyd.

The Evo III was indeed a very advanced vehicle, incorporating a so-called PPC system to reduce turbo lag, and active 4WD. It used Enkei 15-inch rims on gravel events, although up to 18-inch wheels (from the same supplier – not OZ, as on the road cars) were employed on tarmac events. With its new Ohlins dampers and improved aerodynamics, the Evolution III was homologated on 1 April 1995, still using number A5469.

Schwarz went back to Toyota for the 1995 season, so Tommi Mäkinen took his place as a full Mitsubishi works driver. Kenneth Eriksson renewed his Ralliart contract, and the Italian, Andrea Aghini, drove for the team in a number of events. Finally, Phil Short joined the MRE management

The rival teams – Toyota

Like Datsun (Nissan) and Mitsubishi, Toyota was one of the first Japanese makers on the WRC scene. Ove Andersson's Cologne-based Toyota Team Europe used the early Celica and Corolla models in numerous events in the seventies, and from these humble beginnings, TTE blossomed into one of the strongest outfits in rallying. When the third generation Celica gave way to the fourth, the ST165 GT-Fours became regular winners, and the ST185 that followed ultimately gave Toyota its first WRC title in 1993. This was followed by another win in 1994, when the ST205 Celica GT-Four was phased in, and then came a bombshell – a 12-month ban for technical irregularities that meant Toyota was out of the Championship. The Japanese giant eventually made a works team comeback at the end of 1996, but the Corolla WRC would be the new weapon of choice for TTE, making its debut at the 1000 Lakes Rally in August 1997. By 1998, Toyota was back on the pace, posing a real threat to the Mitsubishi and Subaru equipes, and 1999 brought TTE its third title. After this, the company decided to concentrate its efforts on an F1 campaign …

team from TTE. A highly experienced co-driver in his younger days, he took over the position of Team Manager, reporting directly to Andrew Cowan.

With the Lancer Evolution II at his command, Mäkinen was absolutely stunning. He had been lying third for most of the Monte, but trouble with the new differentials on the final day ultimately cost him a place. Aghini went well, finishing sixth, and Holderied/Thorner were the first ladies home, giving them a good start in defence of the previous year's title.

Sweden made a welcome return to the WRC calendar in 1995 and provided Mitsubishi with the venue for a one-two victory in Group A, and a one-two-three in Group N. This was the first time that the Lancer Evolution series had won a World Championship event, but the car had definitely 'come good,' as the Mitsubishis dominated from the very start. Eriksson was leading the event until bad weather set in, so Mäkinen (under team orders) let him through on the final stage to take the victor's laurels.

Mitsubishi led the Championship going into Portugal, but the Group A machines stayed in Rugby. The German team upheld Ralliart honour, though, with two top ten places and another Ladies' prize for Holderied. Thanks to their efforts (which equated to a one-two-three in Group

N), Mitsubishi managed to hold onto its lead at the head of the field.

The classic Safari Rally was not part of the WRC for 1995, although it did count as one of the rounds in the two-litre category. Kenjiro Shinozuka had the car he'd used for the previous two years updated to Evolution III spec, and eventually came home with a well-deserved second place.

The proper Evolution III models made their debut on the Tour de Corse. Aghini showed his mettle on the fast tarmac, coming third in a freshly prepared spare car, while Mäkinen gained valuable experience on an event that was totally new to him. The German Group N machines (still in Evo II guise) took a third successive one-two-three victory!

The WRC circus moved to New Zealand next, where the Ralliart cars were sponsored by Rothmans. Mäkinen led after the first day, but retired on the second after he went off the road. Fortunately, Eriksson managed to stay in touch with the leaders, although his fifth place was not enough to keep Mitsubishi in the number one slot. A good performance from the Celicas gave the Championship lead to Toyota, while McRae's win brought Subaru into contention as well.

Mäkinen won the 1000 Lakes for the second time (he won for Ford in 1994), but it didn't count towards the

WRC. The next round that did count was held in Australia. Mitsubishi again dominated Group N, but Eriksson also brought the Japanese manufacturer its second overall win of the year after the other leading cars suffered varying degrees of misfortune. The Swede remained consistent throughout the event, and was there to reap the reward at the end. Mäkinen was running third for some time but eventually finished fourth, while the Group N cars were once again very strong.

Toyota and Mitsubishi were neck-and-neck going into the next round, the Catalunya in Spain. It was an event filled with controversy, with Toyota caught cheating (the infamous restrictor debacle), and Subaru telling Colin McRae to hand over the win to his Subaru team-mate on the road. However, Holderied retained the FIA World Ladies' Cup (despite not finishing) and fellow Mitsubishi driver, Rui Madeira, secured the Group N title with a Class win. The Germans were also using the Evo III by now, incidentally.

The MRE cars were strangely off the pace in Spain (it was a new event for the team, which goes some way towards explaining this); Aghini won one stage, and collected enough points to put Mitsubishi ahead in the Championship, but Mäkinen made a dramatic exit, skidding off the road on the last day.

Mitsubishi led Subaru by two points going into the RAC – the last round of the season. Fate, however, saw to it that Colin McRae became the youngest ever World Rally Champion, and that Subaru took its first Manufacturers' title. Mäkinen and Eriksson were leading after the first day, but the Finn crashed on the second, damaging his suspension in the process. Meanwhile, Eriksson decided to take his car for a swim, which was rather a sad end to his career with Mitsubishi (he moved to Subaru, having finished third in the Drivers' Championship).

In the increasingly important Asia-Pacific Championship, the Evolution III model took the first two places in Malaysia, a first in Australia (also a round of the WRC), and a one-two on the Hong Kong-Beijing Rally, when Ari Vatanen appeared as a guest. There was another one-two victory in Thailand (the last of six rounds), securing the title for Kenneth Eriksson and Mitsubishi respectively.

1995 RALLY RECORD

No	Driver/co-driver	Position	Reg No (Group)
Monte Carlo (21-26 January)			
11	Tommi Mäkinen/Seppo Harjanne	4th	M5 MRE (Gp A)
12	Andrea Aghini/Sauro Farnocchia	6th	M6 MRE (Gp A)
17	Isolde Holderied/Tina Thorner	10th	GG-R76 (Gp N)
Sweden (10-12 February)			
10	Kenneth Eriksson/Staffan Parmander	1st	K6 MRE (Gp A)
11	Tommi Mäkinen/Seppo Harjanne	2nd	K7 MRE (Gp A)
12	Kenneth Backlund/Tord Andersson	10th	PHB 020 (Gp N)
Portugal (8-10 March)			
12	Rui Madeira/Nuno Silva	9th	GG-R74 (Gp N)
11	Jorge Recalde/Martin Christie	10th	GG-PR5 (Gp N)
10	Isolde Holderied/Tina Thorner	11th	GG-R76 (Gp N)

Continues overleaf

Tour de Corse (3-5 May)

12	Andrea Aghini/Sauro Farnocchia	3rd	K7 MRE (Gp A)
11	Tommi Mäkinen/Seppo Harjanne	8th	M5 MRE (Gp A)

New Zealand (27-30 July)

10	Kenneth Eriksson/Staffan Parmander	5th	M9 MRE (Gp A)
18	Jorge Recalde/Martin Christie	9th	GG-PR5 (Gp N)
17	Rui Madeira/Nuno Silva	10th	GG-R74 (Gp N)
12	Ed Ordynski/Mark Stacey	11th	DJ 5806 (Gp N)
11	Tommi Mäkinen/Seppo Harjanne	dnf	M10 MRE (Gp A)
20	Isolde Holderied/Tina Thorner	dnf	GG-R76 (Gp N)

Australia (15-18 September)

10	Kenneth Eriksson/Staffan Parmander	1st	M7 MRE (Gp A)
11	Tommi Mäkinen/Seppo Harjanne	4th	M8 MRE (Gp A)
12	Ed Ordynski/Mark Stacey	8th	DJ 5806 (Gp N)
15	Jorge Recalde/Martin Christie	10th	GG-PR5 (Gp N)
19	Isolde Holderied/Tina Thorner	19th	GG-R76 (Gp N)
17	Rui Madeira/Nuno Silva	dnf	GG-R74 (Gp N)

Spain (23-25 October)

12	Andrea Aghini/Sauro Farnocchia	5th	M6 MRE (Gp A)
14	Rui Madeira/Nuno Silva	11th	GG-R54 (Gp N)
11	Tommi Mäkinen/Seppo Harjanne	dnf	M5 MRE (Gp A)
17	Jorge Recalde/Martin Christie	dnf	GG-R53 (Gp N)
20	Isolde Holderied/Tina Thorner	dnf	GG-R24 (Gp N)

RAC (19-22 November)

12	Rui Madeira/Nuno Silva	7th	GG-R54 (Gp N)
26	Isolde Holderied/Tina Thorner	14th	GG-R24 (Gp N)
10	Kenneth Eriksson/Staffan Parmander	dnf	K4 MRE (Gp A)
11	Tommi Mäkinen/Seppo Harjanne	dnf	K5 MRE (Gp A)

The 1996 rally season

Tommi Mäkinen was the only driver in the MRE camp to have a full programme, although rookie Brit Richard Burns signed up as the team's number two, mainly to compete in the Asian-Pacific rounds. Former World Champion Didier Auriol made some guest appearances at the end of the season. One wonders what the final 1996 results standing would have been with the talented Frenchman on board for all the events ...

Nonetheless, the Evolution III came good, adding to last year's Australian win with victories in Sweden, Africa (on the Safari Rally), Argentina, Finland and Australia. The same model also took the first three places on the 1996 Hong Kong-Beijing Rally.

Due to the rotation system, the classic Monte was not part of the WRC calendar for 1996, although a privately-entered Group N Lancer managed to take 11th, beating some of the Group A W2L cars in the process.

The WRC season kicked off with the Swedish Rally, with Tommi Mäkinen giving the Evolution III a convincing win. He won ten of the 27 stages, as if to prove a point after last year, when he pulled over to let Eriksson through. The Group N cars dominated their Class, with Backlund winning (14th overall) in his Ralliart Sweden Lancer.

Round two was the Safari. Mäkinen broke a driveshaft on the first day, but still finished it in a strong second place. By the end of day two, he'd moved into the lead, and managed to stay there until the finish in Nairobi. He beat former team-mate Kenneth Eriksson by 14 minutes. As always, Kenjiro Shinozuka was present, coming home in sixth. It was the first time in over a decade that someone had won this gruelling event at the first attempt, but it was

Contemporary advertising linking the Lancer with its race-proven technology.

Richard Burns.

the result of 20 days' practice and a great deal of teamwork. Before the start of the African classic, Phil Short observed: "We've done a lot of things to the car, following on from things that we learnt during the recce. Every problem we had, we managed to rectify."

The WRC then moved to Indonesia. After making a good start, Mäkinen retired with a sick engine on day two. Burns and Yoshihiro Kataoka also went out with engine maladies. It was a disappointing event, giving Subaru the lead in the title chase.

On the Acropolis, Subaru pulled further away after Colin McRae led from start to finish, although Mäkinen (with second) managed to hold on to his lead in the Drivers' Championship. The Mitsubishis came home one-two-three in Group N.

A good result in Argentina (Mäkinen won easily, Richard Burns came fourth, and the Group N cars again claimed the first, second and third slots in their Class) kept Mitsubishi in the title hunt. It also gave the Finn a bigger lead over Carlos Sainz, who had been second in the Drivers' Championship since Indonesia.

The 1000 Lakes Rally was incredibly fast and furious, with Tommi Mäkinen ultimately taking the spoils after an event-long tussle with Juha Kankkunen. Mäkinen therefore extended the gap between him and Sainz, while Mitsubishi moved in front of Subaru.

The WRC then moved to the other side of the world, to a very wet Australia. Mäkinen led for most of the event, and stayed there until the end to secure the Drivers' crown. Mitsubishi also extended its lead, but the Ralliart team was still neck-and-neck with

Subaru – a fact that became clear in San Remo when Mäkinen had a bad accident (on the first stage!) and the other Lancer drivers failed to get on the podium.

In the final round, held in Spain, the Mitsubishi team could do little to stop Subaru claiming its second consecutive title. Mäkinen had a new co-driver (Harjanne had been injured in San Remo) which slowed him down, Burns was off the pace completely, and there was another consequence of the San Remo accident – no car for Didier Auriol, who would have undoubtedly been fast on the

Left and below: 1996 Australian Rally (Mäkinen).

1996 Rally of
Argentina (Mäkinen).

1996 San Remo
Rally (Auriol).

tarmac stages. At least the Lancer driver, Gustavo Trelles, took the Group N crown, narrowly beating Uwe Nittel after a season-long battle.

Tommi Mäkinen won the 1996 WRC Drivers' title – the fifth Flying Finn to do so, but the first with a Mitsubishi. The Lancer had made its mark at the very highest level of competition. Mäkinen had, too, in a season in which he either won the rally, or crashed out trying to do so. Andrew Cowan was heard saying: "It's unbelievable, even to me, how easy he makes it."

1996 Hong Kong-Beijing Rally.

1996 RALLY RECORD

No	Driver/co-driver	Position	Reg No (Group)
Sweden (9-11 February)			
7	Tommi Mäkinen/Seppo Harjanne	1st	M5 MRE (Gp A)
8	Kenneth Backlund/Tord Andersson	14th	HYY 421 (Gp N)
9	Uwe Nittel/Tina Thorner	16th	GG-R54 (Gp N)
Safari (5-7 April)			
7	Tommi Mäkinen/Seppo Harjanne	1st	N4 MRE (Gp A)
8	Kenjiro Shinozuka/Pentti Kuukkala	6th	KNY71-RU5560 (Gp A)
Indonesia (10-12 May)			
32	Chandra Alim/Prihatin Kasiman	9th	BK895RI (Gp N)
21	Bambang Hartono/Agung Baskoro	10th	B9128XX (Gp N)
7	Tommi Mäkinen/Seppo Harjanne	dnf	BK889RI (Gp A)
8	Richard Burns/Robert Reid	dnf	BK890RI (Gp A)
Acropolis (2-4 June)			
7	Tommi Mäkinen/Seppo Harjanne	2nd	M6 MRE (Gp A)
8	Uwe Nittel/Tina Thorner	14th	GG-P47 (Gp N)
Argentina (4-6 July)			
7	Tommi Mäkinen/Seppo Harjanne	1st	N6 MRE (Gp A)
8	Richard Burns/Robert Reid	4th	N7 MRE (Gp A)
9	Uwe Nittel/Tina Thorner	10th	GG-R54 (Gp N)
1000 Lakes (23-26 August)			
7	Tommi Mäkinen/Seppo Harjanne	1st	M5 MRE (Gp A)
8	Lasse Lampi/Jyrki Stenroos	8th	L6 MRE (Gp A)
Australia (15-18 September)			
7	Tommi Mäkinen/Seppo Harjanne	1st	N6 MRE (Gp A)
8	Richard Burns/Robert Reid	5th	N7 MRE (Gp A)
9	Ed Ordynski/Mark Stacey	10th	RV 1404 (Gp N)
San Remo (13-16 October)			
8	Didier Auriol/Denis Giraudet	8th	M7 MRE (Gp A)
7	Tommi Mäkinen/Seppo Harjanne	dnf	M5 MRE (Gp A)
9	Uwe Nittel/Tina Thorner	dnf	GG-Y17 (Gp N)
Spain (4-6 November)			
7	Tommi Mäkinen/Juha Repo	5th	M5 MRE (Gp A)
8	Richard Burns/Robert Reid	dnf	M7 MRE (Gp A)

The 1997 rally season

あなたと創る Creating Together 三菱自動車
シートベルトをしめて、スピードをひかえめに、安全運転は三菱の願いです

ランサーの走り、
ここに極まる。

ランサーの走行性能を極限まで高めて、エボリューションⅣデビュー。

胸のすく走りのランサーをベースに、WRC（世界ラリー選手権）の激しいバトルのなかで磨かれたテクノロジーを集結した
スポーツセダンが、今デビューする。ランサーGSRエボリューションⅣ。耐性と空力特性を高めたボディに
最高出力280PS/6500rpm、最大トルク36.0kg-m/3000rpmを誇るクラス最強のエンジン。さらに動力安全性能を
バックアップし、かつてない旋回・加速能力を引き出す三菱オリジナルのテクノロジー。
AYC（アクティブヨーコントロールシステム）を搭載。スポーツドライブの革命は、ここから始まる。

PHOTO（右）:GSR EVOLUTION Ⅳ ●全長4330mm ●全幅1690mm ●全高1415mm ●ホイールベース2510mm ●主要装備:AYC（アクティブヨーコントロールシステム）、
PIAA社製大型フロントフォグランプ、リヤデスクタ型ウィッカー大型リヤスポイラー、MOMO社製本革巻ステアリングホイール、RECARO社製フルバケットシート、
速転度&ヘデ毒SRSエアバックシステム、4ABS／PHOTO（左）:MX TOURING

LANCER EVOLUTION Ⅳ DEBUT!

Engine:2000 DOHC 16VALVE INTERCOOLER TURBO MaxPower:280PS/6500rpm(NET). MaxTorque:36.0kg-m/3000rpm.

特別低金利キャンペーン実施中
5.8%

A fifth generation Lancer had been launched in Japan back in the autumn of 1995. Although displaying slightly sharper lines than its predecessor, it looked similar to the outgoing model, and dimensions were more or less carried over. However, the body was stronger and lighter, thanks to the help of CAD/CAM technology.

Interestingly, the fifth generation Lancer had been designed from day one with the Evolution IV in mind. For this reason, the engine was still mounted in a transverse position, but 180 degrees different to the earlier models. The new 4WD transmission system sported three shafts instead of two, and it was this third shaft that carried the electromagnetic clutch system which controlled the front and centre differentials on the rally cars. A helical lsd was listed as an option on the sporty grades.

The suspension was very similar to that of the last generation, with a few minor refinements (MacPherson struts up front, with a multi-link rear). Ventilated discs were standard at the front, with either drums or solid discs at the back, depending on the grade. ABS was standard on the top GSR, and optional on most other models; 15-inch alloys were also specified for the 1.8-litre turbocharged

Japanese advertising for the Lancer Evolution IV. Mitsubishi's rally ace Kenjiro Shinozuka is seen standing with the Evo, while the MX Touring in the background has the Japanese actor, Saburo Tokito, giving it his approval.

1997 Tour de Corse (Nittel).

car that would provide the basis for the Lancer GSR Evolution IV.

Announced on 30 July 1996, the all-new Evo IV eventually went on sale on 23 August. Given the CN9A-SNGF code for the GSR (CN9A-SNDF for the RS), the Evolution IV's development was based on Mitsubishi's WRC experience. Due to the demands of the motorsport people, even the styling was based on efficiency rather than cosmetics.

There was a new front bumper (with an integrated grille) and front air dam (with built-in, large diameter PIAA foglights), revised side skirts and rear valance, and a bigger rear spoiler with a delta-shaped wicker, or base; there was also a larger air outlet in the aluminium bonnet.

The overall length was up by 0.79in (20mm), the width and height were 0.20in (5mm) less, and the wheelbase was the same. The front track was increased by 0.20in (5mm), while that at the rear stayed the same. Thus, both were now 57.9in (1470mm).

The coefficient of drag was Cd 0.30, with lift listed as zero. In addition, the body was some 45 per cent stronger than that of the standard Lancers, with extensive additional spot welding, and reinforcements around the scuttle, suspension towers and upper frame; the RS also gained a front tower brace plus additional front crossmember. Not surprisingly, the new car was heavier by around 198lb (90kg); to counteract this extra weight, the Evolution IV was given more power.

The 4G63 Turbo engine was basically the same, although there were many important differences compared to earlier versions. The compression ratio was lowered slightly (from 9.0 to 8.8:1), yet power and torque increased. This was primarily achieved by adopting a new, twin-scroll turbocharger (type TD05HR-16G6-9T) and a 15 per cent

1997 Monte Carlo Rally (Mäkinen).

larger capacity intercooler, although there were a number of other modifications, too.

The cylinder head and lower part of the block were made thinner to reduce weight. The head covered different camshafts (a red rocker cover became a feature at this time, incidentally), whilst the block housed lighter, forged pistons. Between the two was a stainless steel instead of carbon head gasket.

There was a bigger radiator, a lighter flywheel, straighter intake tracts, and a secondary air injection system on the exhaust manifold to reduce exhaust gas interference and keep the turbo spinning hard, even at low revs, thereby reducing lag (this used to be on the rally versions only). One big bore exhaust pipe exited from the back now, instead of twin pipes as found on earlier Evo models and the current production GSR.

Power was now quoted at 280bhp (Japan's maximum, produced at 6500rpm), while maximum torque output went up to 260lbft at 3000rpm. The engine was still transverse, but mounted 180 degrees opposite to that of the Evolution III which, much to the annoyance of the competitions specialists, meant accommodating

1997 Rally of Argentina (Mäkinen).

44

1997 Safari Rally (Burns).

a larger transfer box – it was bigger, heavier and more expensive to produce.

A new W5M51 gearbox with shorter shift strokes was employed, coming with revised ratios (still close), plus the option of a 'low' and 'high' final-drive on the RS. The GSR was listed with 2.785 on first, 1.950 on second, 1.407 on third, 1.031 on fourth and 0.761 on top, while RS had the same bottom two ratios, but a closer 1.444 on third, 1.096 on fourth and 0.825 on fifth. The final-drive was 4.529:1 on the GSR and RS 'high' option, or 4.875 on the RS's 'low' specification.

A new feature was the AYC (Active Yaw Control System) rear differential, which used electronics to hydraulically give more torque to the outside wheels, and less to the inside one to improve cornering. The GSR came with AYC at the rear, a viscous-coupled centre differential with a 50/50 split, and a helical front lsd; the RS's rear lsd was a 1.5-way mechanical type, while the torque-sensing helical front was listed as an option.

1997 1000 Lakes Rally (Mäkinen).

45

1997 Rally of Portugal (Mäkinen).

There was a new multi-link rear suspension, and revised geometry up front to give a lower roll centre; anti-roll bar diameters were 0.90in (23mm) up front, and 0.83in (21mm) at the back.

Brake feel was enhanced through the use of bigger ventilated discs on the GSR – 11.6in (294mm) at the front, and 11.2in (284mm) at the rear – possible because of the move to 6.5J x 16 OZ 12-spoke alloys with 205/50 VR16 tyres. Interestingly, the rear discs incorporated a small drum for a more efficient handbrake.

ABS was standard, although the RS – still built to order – ran on 6.5J x 15 steel rims shod with HR-rated rubber,

and had the smaller diameter brakes carried over from the Evo III; the 16-inch wheel and tyre combination was an option on the RS. In fact, there were five RS set packages: AYC, 16-inch wheels and tyres, and bigger brakes; a front helical lsd, 16-inch wheels and tyres, and bigger brakes; a close-ratio gearbox with a 'high' final-drive (for circuit use), a front helical lsd, 16-inch wheels and tyres, and bigger brakes; a close-ratio gearbox with a 'low' final-drive (for rallying and similar disciplines) plus the front helical lsd, and, finally, the latter option with 16-inch wheels and tyres, and bigger brakes.

A total of 6000 Evo IVs were scheduled to be built,

1997 Rally of New Zealand (Burns).

with a mix of GSR and RS grades as always, and just over 100 were exported for competition users. However, the Evo IV wouldn't make its works debut until the start of the 1997 rally season, with Tommi Mäkinen as the main MRE driver.

The controversial 'event rotation' system was no longer employed, so the 1997 season took in a total of 14 rounds (to qualify, manufacturers had to enter all of them). In addition, the FIA's new World Rally Car rules allowed a touch more freedom with vehicle design, in that the minimum 2500 homologation run was no longer necessary. Specialist machines could be built by outside concerns, as long as they were based on a production model.

Testing of the Evolution IV had started in the latter part of 1996, although the Evo III continued to be the works car for the entire 1996 campaign. Unlike earlier models in the Evolution series, it was not possible to upgrade an earlier vehicle with a later specification, so the 1997 cars were completely new machines. Homologated on 1 January 1997, the Evolution IV featured a sequential gearbox, and water-cooled brakes for the faster tarmac events. However, of the front runners, only Mitsubishi failed to take full advantage of the new World Rally Car rules, deciding to run traditional Group A cars instead.

With a full 14-round season, the Monte made a

welcome return to the WRC calendar. It was an interesting event, with four different leaders on each of the four days. Sadly, although Mäkinen led on the third day, Subaru's Piero Liatti finished ahead on the one that really mattered – the final day! Uwe Nittel did well as the other nominated driver (in an old Evo III), while Gustavo Trelles walked away with Group N honours, coming home ninth overall.

Mäkinen had bad luck in Sweden, when problems occurred after a routine gearbox change. The delay in curing the trouble cost him time, and he did well to fight back to third place. Nittel had an accident on the last day, although Kenneth Bäcklund took a Group N Class win (11th overall).

The Safari was not kind to Mäkinen, with suspension and transmission problems and a series of punctures ultimately putting him out of the event. Burns did well in the 'Carisma GT' (an Evo IV with different badges), but it was the only Mitsubishi to finish in the top ten.

At last, Mitsubishi managed to break Subaru's winning run with a victory in Portugal. It was the first win for the Evo IV, and the first WRC win for any car equipped with a sequential gearbox. Mäkinen was full of praise for the innovative transmission, which gave him a four minute

1997 Australian Rally (Mäkinen).

lead over his nearest rival when the dusty event came to an end. Trelles won Group N, coming seventh overall, and his Ralliart Germany team-mate, Manfred Stohl, was eighth.

Mäkinen kept up the pressure in Spain with his first victory on a tarmac rally. He ran a consistent rally, as did Trelles, who took his third Group N win of the season. On the Tour de Corse, however, it was a different story, with the Flying Finn going over the side of a cliff after hitting a stray cow! It was not a good event for Mitsubishi, allowing Subaru to pull further ahead in the title chase.

Argentina marked the debut of the Evolution IV in Group N trim, but failed to perform well after a pre-start mishap. However, Trelles won the Class (sixth overall) in his older model and, more importantly, Tommi Mäkinen held up the winner's trophy at the finish in Cordoba. He had led almost from the start, claiming the best time on ten of the 23 stages.

Tyre trouble delayed progress somewhat on the Acropolis, almost certainly costing Mäkinen victory. And there was more sadness in the Mitsubishi camp after Akira Kondo, Ralliart's President, succumbed to a long illness. He left quite a legacy, though, which was now taken care of by Taizo Yokoyama (the new President). Yokoyama was assisted by Kiyohito Iritani, who was given the General Manager's post.

As attention turned to New Zealand, things didn't get much better. Mäkinen crashed out, making his Lancer into a hatchback, and Burns had transmission maladies. At least the Finn held on to his lead in the Drivers' Championship, and Trelles was on hand to win Group N again, coming seventh overall in GG-R73.

In his home event, Mäkinen reigned supreme. This was his fourth straight win in Finland: he had won for Ford in 1994, for Mitsubishi in 1995 (when it didn't count towards the WRC), 1996 and 1997. With Sainz and McRae falling by the wayside, Mäkinen managed to extend his lead, and Trelles claimed the Group N title with four events still to go.

It was a different story in Indonesia, though, with Mäkinen retiring after an event full of drama. Despite not finishing, Mäkinen held on to his lead, though the margin

1997 RAC Rally (Mäkinen).

between him and Sainz (who won the rally) was now only eight points.

There had been changes in the MRE staff at the start of the season (a key engineer, Marc Amblard, left the team, although his place was quickly taken by ex-Ford man, Bernard Lindauer), but the news of Roland Lloyd going was a major blow. The highly respected Chief Engineer joined FFD Ricardo in the autumn of 1997. The team suffered another setback when Seppo Harjanne announced his retirement at the end of the season. He took his place in history as one of the most successful co-drivers ever, but it left Tommi Mäkinen with an empty passenger seat. Even after finding a new top class co-driver, it would doubtless take him time to build up the rapport he had enjoyed

The rival teams – Subaru

Subaru's first real efforts in the WRC came via the Group A Legacy RS, which ultimately gave way to the 4WD Impreza in 1993. By this time the works team was being run by Prodrive in England, and the results spoke for themselves – the Group A machine took the WRC crown in 1995 and 1996, while a new WRCar model claimed the title in 1997. The Impreza continued to be a threat to all comers in the years that followed, although there were to be no more titles, and Subaru duly announced its withdrawal from rallying at the end of 2008.

with Harjanne. At least Mäkinen had a few more events alongside his fellow Finn.

Unsuitable suspension settings held up Mäkinen in San Remo, and Nittel, who had been so consistent throughout the season, had a series of accidents that eventually put an end to his rally. The only saving grace was that Sainz came fourth, thus leaving Mäkinen ahead in the Drivers' Championship.

There was more trouble in Australia, but this time it was Mäkinen's fault – he paid the consequences of a heavy landing and an argument with a tree, but still, somehow, managed to finish second, leaving him only one point to get in order to secure his second Championship, assuming Colin McRae – the man in second place after his victory in Oz – won the RAC. (McRae's win gave Subaru the Manufacturers' title, incidentally.) Local man, Ed Ordynski,

came sixth on the event, while an Australian Mitsubishi crew took the Ladies' Prize.

Fending off a fever, Tommi Mäkinen did just enough on the RAC. Although he didn't even win a stage, he came home in sixth, two places behind the 'Carisma GT' of Burns. McRae did win the event with an impressive display of car control, which meant a nail-biting finish to a closely-fought season.

Although the season was not the best for Mitsubishi, at least Tommi Mäkinen held onto his Drivers' Championship, narrowly beating Colin McRae (by just one point!). Gustavo Trelles, however, had a much easier time in the Group N category – he also managed to retain his title, having more than three times the points of his nearest competitor. Subaru was declared Champion again (its third consecutive win), with Ford second and Mitsubishi third.

1997 RALLY RECORD

No	Driver/co-driver	Position	Reg No (Group)
Monte Carlo (19-22 January)			
1	Tommi Mäkinen/Seppo Harjanne	3rd	P22 MRE (Gp A)
2	Uwe Nittel/Tina Thorner	5th	N7 MRE (Gp A)
Sweden (7-10 February)			
1	Tommi Mäkinen/Seppo Harjanne	3rd	P2 MRE (Gp A)
2	Uwe Nittel/Tina Thorner	dnf	N7 MRE (Gp A)
Safari (1-3 March)			
2	Richard Burns/Robert Reid	2nd	P33 MRE (Gp A)
1	Tommi Mäkinen/Seppo Harjanne	dnf	P3 MRE (Gp A)

Portugal (23-26 March)			
1	Tommi Mäkinen/Seppo Harjanne	1st	P22 MRE (Gp A)
2	Richard Burns/Robert Reid	dnf	P2 MRE (Gp A)
Spain (14-16 April)			
1	Tommi Mäkinen/Seppo Harjanne	1st	P44 MRE (Gp A)
2	Uwe Nittel/Tina Thorner	8th	N7 MRE (Gp A)
Tour de Corse (5-7 May)			
2	Uwe Nittel/Tina Thorner	8th	N7 MRE (Gp A)
1	Tommi Mäkinen/Seppo Harjanne	dnf	P44 MRE (Gp A)
Argentina (22-24 May)			
1	Tommi Mäkinen/Seppo Harjanne	1st	P4 MRE (Gp A)
2	Richard Burns/Robert Reid	dnf	P5 MRE (Gp A)
Acropolis (8-10 June)			
1	Tommi Mäkinen/Seppo Harjanne	3rd	P22 MRE (Gp A)
2	Richard Burns/Robert Reid	4th	P2 MRE (Gp A)
10	Uwe Nittel/Tina Thorner	6th	N7 MRE (Gp A)
New Zealand (2-5 August)			
2	Richard Burns/Robert Reid	4th	P5 MRE (Gp A)
1	Tommi Mäkinen/Seppo Harjanne	dnf	P4 MRE (Gp A)
1000 Lakes (29-31 August)			
1	Tommi Mäkinen/Seppo Harjanne	1st	P55 MRE (Gp A)
2	Uwe Nittel/Tina Thorner	7th	N7 MRE (Gp A)
Indonesia (19-21 September)			
2	Richard Burns/Robert Reid	4th	P33 MRE (Gp A)
1	Tommi Mäkinen/Seppo Harjanne	dnf	P3 MRE (Gp A)
San Remo (12-15 October)			
1	Tommi Mäkinen/Seppo Harjanne	3rd	P44 MRE (Gp A)
2	Uwe Nittel/Tina Thorner	dnf	N7 MRE (Gp A)
Australia (30 October-2 November)			
1	Tommi Mäkinen/Seppo Harjanne	2nd	P5 MRE (Gp A)
2	Richard Burns/Robert Reid	4th	P33 MRE (Gp A)
RAC (23-25 November)			
2	Richard Burns/Robert Reid	4th	P55 MRE (Gp A)
1	Tommi Mäkinen/Seppo Harjanne	6th	P6 MRE (Gp A)

The 1998 rally season

Visitors to the 1997 Tokyo Show got a sneak preview of the Type CP9A Lancer Evolution V. Although it still had 280bhp, there was a substantial increase in torque output, and its attractive, 17-inch, aluminium alloy wheels concealed powerful Brembo brakes. Press material handed out at the event said that it would be available in the spring, whetting the appetite of motorsport enthusiasts everywhere. Ultimately announced on 6 January 1998, sales began just three weeks later.

The aluminium bonnet design was revised for better heat dissipation, and the same lightweight material was adopted for the flared front wings (the rear fender size was increased via wheelarch blisters). There were new front and rear bumpers, a different air dam, modified side and rear skirts, and a four-position rear spoiler with a delta-shaped wicker and aluminium wing.

1998 San Remo Rally (Mäkinen).

1998 RAC Rally (Mäkinen).

1998 Catalunya Rally (Mäkinen).

As for the leading dimensions, while the height (at 55.7in, or 1415mm) and wheelbase (at 98.8in, or 2510mm) remained the same, the length was now 171.2in (4350mm), which represented an increase of 0.79in (20mm), and the width was increased by 3.15in (80mm), taking it to 69.7in (1770mm) overall. The track measurements were also wider: 59.4in (1510mm) at the front, and 59.2in (1505mm) at the back (increases of 1.57 and 1.38in, or 40 and 35mm respectively). The bigger car followed the trend started by World Rally Car regulations, although the Evolution V still managed to just keep within the Group A regulations. The Cd was quoted as 0.31, with zero lift.

The engine was basically the same, with 8.8:1 compression ratio retained. However, there was a modified twin-scroll turbocharger (type TD05HR-16G6-10.5T) and intercooler, new lightweight pistons, and the radiator and oil cooler capacities were increased. Power stayed at the Japanese maximum of 280bhp (produced at 6500rpm), but torque was enhanced somewhat – now listed at 274lbft at 3000rpm.

Gear ratios were carried over from the Evo IV (including the options on RS), but the synchromesh and shift linkage was made stronger, topped with a smaller gearknob trimmed in black leather with red stitching (actually, a number of road tests mentioned the better gearshift). The innovative AYC system continued on the GSR, matched with a helical lsd up front.

There was a longer lower arm on the front suspension, made from forged aluminium alloy, and the inverted front struts were given longer strokes. Mountings were changed at the rear to give the car its wider track, while revising the geometry at the back end gave a lower roll centre, thus enhancing roadholding and vehicle response during cornering. In addition, the steering rack location was altered, along with the knuckle joint location, to give more linear response in corners, while a new pump was adopted, allowing the engineers to delete the power steering oil cooler, and thus save weight.

Tyres were now 225/45 ZR17 mounted on OZ alloys – although 7.5J x 17, they were nonetheless of similar design to those fitted on the Evo IV. As a result, the front gained four-pot calipers (formally two-pot) and bigger 12.6in (320mm) diameter discs, while two-pot calipers were employed at the back (earlier Evolution models used to have single cylinder calipers), again with larger 11.8in (300mm) diameter discs. Brakes were made by Brembo, with ABS coming as standard on the GSR.

For the first part of the 1998 WRC season, there was little change in the Ralliart camp – the same car, and the same driver line-up. All of the leading cars were listed with a power output of around 300bhp, weighing in at

1998 RAC Rally (Burns).

1998 Australian Rally (Mäkinen).

2706lb (1230kg). Mitsubishi quoted 300bhp at 6000rpm and 375lbft of torque at 3500rpm, both for Mäkinen's Lancer and Burns' so-called 'Carisma GT.' The Evolution IV ultimately won in Sweden and Africa (on the Safari Rally) before it was replaced by a new machine.

Testing of the Evolution V started during the summer of 1997; Lasse Lampi was the chief test driver, helped by Richard Burns. Again, the latter's car was run with Carisma GT badges for marketing reasons. Having made its debut in Spain, the Evolution V won in Argentina, Finland, San Remo, Australia and Great Britain (the RAC Rally). With the victories clocked up by the Evo IV earlier in the season, this was enough to secure the WRC Manufacturers' title for Mitsubishi.

Looking back on the season in detail, Mäkinen had been leading in Monte Carlo until a big accident on the second day. This was Burns' first Monte, so his fifth place was a remarkable achievement, especially given the awful weather. Uwe Nittel came home seventh, while Manfred Stohl took Group N honours.

In Sweden, Thomas Rådström was leading for TTE until halfway, when Mäkinen moved into the number one slot following the Swede's retirement. He managed to stay there until the end. Burns had an excursion on stage two, costing him his chances of a decent finish. Mitsubishi dominated Group N, as had become the norm.

Mäkinen and Burns were running one-two for most of the Safari. However, after a great deal of punishment, the engine mountings gave way on the Finn's car at the end of the second day. (In fact, only 19 of the 48 cars that started managed to complete the event.) Burns took a fine victory, with Luis Climent winning Group N (seventh overall).

In Portugal, Mäkinen hit a tree, ending his rally early on the second day. Burns made a wrong tyre choice towards the end of the event, dashing any hopes of a podium finish, but fourth place was a good, solid result. Gustavo Trelles won the Group N battle.

The Spanish round saw the debut of the Evolution V. Mäkinen's insufficient testing of the new model showed in the early stages, but his natural talent showed through in the end – he finished third behind two Toyotas. Burns wasn't far behind in fourth, and Trelles took Group N again.

Burns had a mishap at the end of the second day in Corsica and his front suspension was too damaged to go on. Meanwhile, Mäkinen had electrical problems, caused by water getting onto the ECU. At least Stohl lifted the trophy for the Group N category.

Mäkinen and Colin McRae had a memorable battle in Argentina, although suspension troubles eventually put the Scot out of contention. Mäkinen went on to win, with Burns in fourth (behind a Toyota and Ford). Trelles once again claimed the Group N Class.

In Greece, electrical problems robbed Mäkinen of any chance of winning, or even finishing in a decent

1998 Rally of Argentina (Mäkinen).

1998 Swedish Rally (Mäkinen).

1998 Safari
Rally (Burns).

1998 1000 Lakes Rally (Mäkinen).

position, so he retired on the second stage. Burns was running in the top five until suspension failure put him out. At least Trelles was there to pick up Group N.

Burns was doing well in NZ until stage 19 (of 25), when he fell from third to 15th. He clawed his way back to ninth before the end. Mäkinen, meanwhile, put in a fairly mediocre performance, but still managed third, although he was a long way down on the winning Toyota. Subaru won Group N, breaking Mitsubishi's incredible run in that category.

With a convincing victory in Finland, Mäkinen set a new record of five consecutive wins on a single rally. It was Burns' first 1000 Lakes, so his fifth position was commendable on this demanding, high speed event. Trelles won Group N.

At San Remo, the rear differential was also equipped with an active electromagnetic clutch to match those already employed up front and in the centre differential. With the help of the latest computer programmes, the 4WD systems could react to the road surface and driver's commands at lightning speed. Mäkinen used this new technology to good effect – he led from stage two all the way to the finish. Burns was never really on the pace, but still managed to claim seventh position.

In Australia, Burns was leading until midway through the second day, when he was overhauled by Sainz (the Championship leader going into the event, albeit by two points), Mäkinen and Auriol. There was drama when McRae moved into the lead with two stages to go, although his turbo blew towards the end, costing him his chances of the WRC crown. Burns had an accident on the final stage, but Mäkinen won for Mitsubishi. Sainz came second to keep the title chase bubbling ...

The Drivers' Championship would be decided once again after the final round, with Mäkinen leading Carlos Sainz by just two points going into it. The Finn had all but given up hope of securing his third consecutive title when he lost a rear wheel following an accident early in the event. However, fate intervened and the engine of Sainz's Toyota blew up just before the finish, handing the crown to Mäkinen. Burns scored a very impressive win, beating Juha Kankkunen by almost four minutes, while Stohl won Group N (tenth overall).

Ultimately, despite the nail-biting drama in the British forests, Mäkinen gained his third consecutive title, finishing the season on 58 points, two ahead of Carlos Sainz, and 13 ahead of Colin McRae. In addition, Mitsubishi lifted the Manufacturers' trophy with 91 points to Toyota's 85 and Subaru's 65, while Gustavo Trelles took the Group N title, handsomely beating fellow Lancer driver, Manfred Stohl.

In fact, Ralliart's only disappointment came in the Asia-Pacific Championship, which had lost some of its sparkle in any case. The FIA stated that no works drivers would be allowed to score points, so there was little point in fielding top runners from the WRC. As such, Mitsubishi pinned its hopes on Yoshihiro Kataoka, who was destined to finish the season in second place behind Toyota's Yoshio Fujimoto.

No	Driver/co-driver	Position	Reg No (Group)
Monte Carlo (18-21 January)			
2	Richard Burns/Robert Reid	5th	P44 MRE (Gp A)
1	Tommi Mäkinen/Risto Mannisenmäki	dnf	P22 MRE (Gp A)
Sweden (5-8 February)			
1	Tommi Mäkinen/Risto Mannisenmäki	1st	P6 MRE (Gp A)
2	Richard Burns/Robert Reid	15th	P55 MRE (Gp A)
Safari (27 February-2 March)			
2	Richard Burns/Robert Reid	1st	P33 MRE (Gp A)
1	Tommi Mäkinen/Risto Mannisenmäki	dnf	P3 MRE (Gp A)
Portugal (22-25 March)			
2	Richard Burns/Robert Reid	4th	P55 MRE (Gp A)
1	Tommi Mäkinen/Risto Mannisenmäki	dnf	P6 MRE (Gp A)
Spain (19-22 April)			
1	Tommi Mäkinen/Risto Mannisenmäki	3rd	R2 MRE (Gp A)
2	Richard Burns/Robert Reid	4th	R22 MRE (Gp A)
Tour de Corse (3-6 May)			
1	Tommi Mäkinen/Risto Mannisenmäki	dnf	R2 MRE (Gp A)
2	Richard Burns/Robert Reid	dnf	R22 MRE (Gp A)
Argentina (20-23 May)			
1	Tommi Mäkinen/Risto Mannisenmäki	1st	R3 MRE (Gp A)
2	Richard Burns/Robert Reid	4th	R33 MRE (Gp A)
Acropolis (5-9 June)			
1	Tommi Mäkinen/Risto Mannisenmäki	dnf	R4 MRE (Gp A)
2	Richard Burns/Robert Reid	dnf	R44 MRE (Gp A)
New Zealand (25-28 July)			
1	Tommi Mäkinen/Risto Mannisenmäki	3rd	R3 MRE (Gp A)
2	Richard Burns/Robert Reid	9th	R33 MRE (Gp A)
1000 Lakes (21-23 August)			
1	Tommi Mäkinen/Risto Mannisenmäki	1st	R66 MRE (Gp A)
2	Richard Burns/Robert Reid	5th	R4 MRE (Gp A)

No	Driver/co-driver	Position	Reg No (Group)
San Remo (10-14 October)			
1	Tommi Mäkinen/Risto Mannisenmäki	1st	R8 MRE (Gp A)
2	Richard Burns/Robert Reid	7th	R2 MRE (Gp A)
Australia (5-8 November)			
1	Tommi Mäkinen/Risto Mannisenmäki	1st	R3 MRE (Gp A)
2	Richard Burns/Robert Reid	dnf	R33 MRE (Gp A)
RAC (21-23 November)			
2	Richard Burns/Robert Reid	1st	R4 MRE (Gp A)
1	Tommi Mäkinen/Risto Mannisenmäki	dnf	R66 MRE (Gp A)

The 1999 rally season

Announced in January 1999, chassis codes, weights and dimensions for the Evolution VI were the same as those listed for the Evo V. The engine and gearbox were basically carried over (although there were a few subtle modifications and a new sump was adopted). There were new OZ wheels, but they wore the same rubber, while the RS version continued on 15-inch rims. In reality, the main changes centred around revised aerodynamic appendages in order for the car to comply with the latest WRC regulations.

The Evo VI featured a new front bumper with

1999 Monte Carlo Rally (Mäkinen).

integrated grille (to meet the 1999 FIA regulations), separate oil and brake cooling ducts (the oil cooler duct was on the offside only, just ahead of the wheel), smaller foglights, and an offset number plate to increase airflow to radiators; the front indicators were now clear. Moving around the vehicle, the old side skirts were retained, but there was a new, smaller rear spoiler with twin blades (it was made smaller to comply with FIA guidelines); beneath the spoiler, the rear light cluster garnish had gone. Ultimately, the Evo VI displayed a Cd figure of 0.30 and zero lift, plus enhanced high speed stability.

Meanwhile, more spot welding and special adhesives were employed to further strengthen the body, with stronger front shock mounting points. The front and rear suspension underwent subtle revisions, with a lower roll centre, and an increased stroke and more forged aluminium parts for the rear. However, the Evo V suspension could still be specified on the RS for those who entered gymkhana-type events.

As for the engine, a larger air intake hose was fitted, and there was better breathing on the turbocharger (the GSR turbo was the same, but the RS had a type TD05HRA-16G6-10.5T unit with a

Freddy Loix.

The twin blades on the rear spoiler brought complaints from rival manufacturers, so the lower wing was made into what equated to a tall wicker via carbonfibre inserts riveted in place.

more responsive titanium-aluminium alloy turbine blade). A bigger oil cooler gave 23 per cent better heat dissipation; the cooling system was also modified, and the lightweight pistons now incorporated oil cooling channels.

A twin-plate clutch was listed as an option for the RS (an hydraulically-operated, single-plate unit was the norm), while the AYC system was improved via rally experience. The Brembo brakes (standard on the GSR, optional on the RS) were carried over, although revised caliper shapes helped to increase strength. Ventilated at both front and back, the diameters were 12.6 and 11.8in (320 and 300mm) respectively.

Mitsubishi's works 1999 WRC campaign employed the Evo VI which, thanks to the sterling work of Bernard Lindauer and his team of engineers, Tommi Mäkinen described as being stable in all conditions. It had the same power and torque as its predecessor, and was also the same weight.

The number two car was again described as a

Carisma GT, but it was driven by Freddy Loix, as Burns had moved to Subaru. Born in 1970, the young Belgian had made quite a name for himself with Toyota (in both the Celica and later the Corolla WRC).

Mäkinen led the Monte Carlo Rally almost from start to finish, while Loix crashed out on the first day – not the best of starts to his Ralliart career! Marc Duez won the Group N category in an Evolution V.

In Sweden, Loix had more than his fair share of trouble, but Mäkinen was once again in top form, beating Carlos Sainz to give him a clear 13 point margin over his nearest rivals in the Championship.

The Safari was full of drama. Mäkinen was excluded after receiving help from spectators whilst changing a tyre. It was an annoying situation, as he had finished second on the road. Loix failed to impress again, crashing out on the first day after having taken a hump far too quickly. Injuries forced him to miss the next round.

Marcus Grönholm, who would later make quite an impression on the WRC scene, was going well in

1999 Rally of Portugal (Grönholm).

Portugal until his clutch gave up. A faulty front differential reduced Mäkinen's fighting potential early on, although he ultimately recovered to take fifth.

New brakes clashing with the transmission's electronics slowed Mäkinen at the start of the Spanish round, but the system was refined as the rally progressed. Later, a jump start cost the Finn a one minute penalty and, in effect, second place. However, Burns received the same punishment, promoting Loix one position, while Trelles won Group N.

Going into the Tour de Corse, Mäkinen led Auriol by three points in the Championship, with McRae third. Toyota led Mitsubishi by 11, with Ford a further four

behind. A good result was needed, but unfortunately both Lancers had brake trouble. At least Trelles won Group N.

As the circus moved to Argentina, things failed to improve. Mäkinen had gearbox troubles that cost him dearly – he was never able to make up the time, although fourth was more than respectable given the circumstances. Loix crashed out, but Trelles left his mark in the Group N Class once again. Both works drivers put in a good solid performance on the Acropolis, resulting in third and fourth. Mäkinen still led the Championship, but Toyota pulled further away.

Tommi Mäkinen won easily in New Zealand, but Loix's eighth place made Ralliart realise it was time to set

1999 Australian Rally
(Mäkinen).

1999 Rally of New Zealand
(Mäkinen).

Japanese advertising from the Enkei concern, celebrating the fact that it had supplied the wheels for both Formula One and WRC Champions for two years running.

up the car for his very different driving style to that of the number one driver, as Trelles came home in ninth in his Group N Evolution V. Unique settings for the Belgian certainly brought about better results.

Mäkinen's hopes of a sixth straight win on the 1000 Lakes came to an end after his car came to a halt with transmission problems, although he had never really troubled Kankkunen. Loix came tenth despite an argument with the Finnish scenery.

China is best forgotten: both works cars had accidents, and Trelles came second in Group N (eighth overall), beaten by the Subaru of Toshihiro Arai – more disappointment. Auriol now tied with Mäkinen in the title chase, while Toyota was running away with the Championship.

In San Remo, Mäkinen came from behind on the final stage to snatch the winner's trophy, and Loix did well to take fourth, the fast tarmac roads suiting his style better than gravel or snow. Burns won in Australia, followed by Sainz. Mäkinen claimed third

1999 Safari Rally (Mäkinen).

1999 San Remo Rally (Mäkinen).

1999 RAC Rally (Loix).

1999 Catalunya Rally (Mäkinen).

1999 Rally of Argentina (Mäkinen).

to keep him ahead in the Championship and, with Auriol retiring early on, it almost secured the title.

Mäkinen went out of the RAC with engine failure, although Didier Auriol went out on the next stage, thus deciding the Championship; ultimately, Burns' charge at the end of the season put him ahead of the TTE driver. Loix came fifth to give him eighth in the WRC.

Mäkinen had done enough to secure his fourth straight Drivers' Championship, scoring 62 points. Burns, his old team-mate, got 55, while Didier Auriol finished the season on 52; Kankkunen and Sainz came next on 44.

Lancer driver Gustavo Trelles took Group N honours again, although the Manufacturers' title went to Toyota (109 points), followed home by Subaru (105), Mitsubishi (83) and Ford (37). In other words, the WRC was totally dominated by Japanese marques.

The rival teams – Peugeot

After gaining two WRC titles in the mid-eighties with the 205 Turbo 16, in an era dominated by Audi and Lancia, Peugeot's efforts on the rally front sadly went quiet – after all, a lot of effort (and money!) had been put into developing Group B machines, and now they were suddenly made redundant. However, Peugeot Sport returned to the fold in 1999 with the 206 WRC, and duly won the World Rally Championship in 2000, 2001 and 2002. Sister company Citroën then started to win everything, despite Peugeot releasing a new 307 WRC model for the 2004 season. Soon after, at the end of 2005, Peugeot left the rallying scene for good.

1999 RALLY RECORD

No	Driver/co-driver	Position	Reg No (Group)
Monte Carlo (17-21 January)			
1	Tommi Mäkinen/Risto Mannisenmäki	1st	S2 TMR (Gp A)
2	Freddy Loix/Sven Smeets	dnf	S22 TMR (Gp A)
Sweden (11-14 February)			
1	Tommi Mäkinen/Risto Mannisenmäki	1st	S3 TMR (Gp A)
2	Freddy Loix/Sven Smeets	9th	S33 TMR (Gp A)
Safari (25-28 February)			
2	Freddy Loix/Sven Smeets	dnf	S44 TMR (Gp A)
1	Tommi Mäkinen/Risto Mannisenmäki	exc	S4 TMR (Gp A)
Portugal (23-26 March)			
1	Tommi Mäkinen/Risto Mannisenmäki	5th	S3 TMR (Gp A)
2	Marcus Grönholm/Timo Rautiainen	dnf	S33 TMR (Gp A)
Spain (19-21 April)			
1	Tommi Mäkinen/Risto Mannisenmäki	3rd	S2 TMR (Gp A)
2	Freddy Loix/Sven Smeets	4th	S22 TMR (Gp A)

No	Driver/co-driver	Position	Reg No (Group)
Tour de Corse (7-9 May)			
1	Tommi Mäkinen/Risto Mannisenmäki	6th	S2 TMR (Gp A)
2	Freddy Loix/Sven Smeets	8th	S22 TMR (Gp A)
Argentina (22-25 May)			
1	Tommi Mäkinen/Risto Mannisenmäki	4th	T2 MMR (Gp A)
2	Freddy Loix/Sven Smeets	dnf	T22 MMR (Gp A)
Acropolis (6-9 June)			
1	Tommi Mäkinen/Risto Mannisenmäki	3rd	S4 TMR (Gp A)
2	Freddy Loix/Sven Smeets	4th	S44 TMR (Gp A)
New Zealand (15-18 July)			
1	Tommi Mäkinen/Risto Mannisenmäki	1st	S2 TMR (Gp A)
2	Freddy Loix/Sven Smeets	8th	T2 MMR (Gp A)
1000 Lakes (20-22 August)			
2	Freddy Loix/Sven Smeets	10th	S33 TMR (Gp A)
1	Tommi Mäkinen/Risto Mannisenmäki	dnf	S3 TMR (Gp A)
China (16-19 September)			
1	Tommi Mäkinen/Risto Mannisenmäki	dnf	-001 (Gp A)
2	Freddy Loix/Sven Smeets	dnf	-002 (Gp A)
San Remo (11-13 October)			
1	Tommi Mäkinen/Risto Mannisenmäki	1st	V2 MMR (Gp A)
2	Freddy Loix/Sven Smeets	4th	V22 MMR (Gp A)
Australia (4-7 November)			
1	Tommi Mäkinen/Risto Mannisenmäki	3rd	S2 TMR (Gp A)
2	Freddy Loix/Sven Smeets	4th	T2 MMR (Gp A)
RAC (21-23 November)			
2	Freddy Loix/Sven Smeets	5th	S33 TMR (Gp A)
1	Tommi Mäkinen/Risto Mannisenmäki	dnf	S3 TMR (Gp A)

The 2000 rally season

The 1999 Tokyo Show saw the debut of the Mitsubishi Lancer Evolution VI Tommi Mäkinen Edition. Officially announced at the end of the year, it was introduced to celebrate Mäkinen's four consecutive World Championship for Drivers titles, and is sometimes referred to as the Evolution VI½.

Engine and transmission options were carried over, along with the gear ratios on the five-speed 'box. However, the high response titanium-aluminium alloy turbine blades were now specified on the GSR's turbocharger as well (type TD05HRA-15GK2-10.5T), combined with a smaller diameter compressor wheel; the RS kept its old turbo, with the new one as an option. In addition, there was a new exhaust with a big bore tailpipe. Ultimately, the engine provided lots of torque in the low- to mid-range, the Evolution VI½ modifications bringing in the maximum torque 250rpm lower down the rev-band.

Easily distinguished by its redesigned, aggressive-looking front bumper/air dam, the Tommi Mäkinen Edition rode 0.39in (10mm) lower than the Evolution VI, as the GSR came with tarmac suspension settings (an option on the RS). A front tower bar was standard on all cars, including the GSR, while the steering ratio was quicker than that of the earlier Evos.

White Enkei 17-inch alloys (the same as those used on the works Group A cars) came as part of the GSR package, and could be bought as an option for the RS grade. Combined with the so-called Special Colour Package (available on the Passion Red GSR as a 20,000 yen option), the exterior could be made to resemble an authentic WRC machine.

The works Evolution VI had 290bhp at 6000rpm and a massive 375lbft of torque at 3500rpm for the 2000 season. Freddy Loix was once again driving a 'Carisma GT,' although, in reality, it was an Evolution VI, almost identical to Mäkinen's car.

The Flying Finn won in Monte Carlo after a masterful performance; he led from the fourth stage and never looked back. Loix was never really happy, but came home in sixth place nonetheless. Manfred Stohl came ninth in an Evolution VI to take Group N honours. However, in Sweden, Mäkinen was beaten by Grönholm, who'd been a guest driver for Ralliart last year when Loix was injured. Loix himself was never on the pace, coming eighth.

On the Safari, the battery packed up after the suspension broke on Mäkinen's car, while Loix went out a couple of stages later with his own suspension maladies. At least Claudio Menzi, driving an Evolution V, took Group N (ninth overall – an excellent result on his first African sortie), to uphold Mitsubishi's unbeaten record in that category (a Lancer had won the Class in Sweden, too).

In Portugal, a broken suspension caused by a crash put Mäkinen out, but he was never higher than eighth in any case. Loix came sixth, but Burns' victory put him ahead in the Drivers' Championship. Mitsubishi's domination in Group N continued, with Miguel Campos taking the spoils in an Evo V.

Moving across the border to Spain brought better fortune. Mäkinen had differential problems, but overcame them, fighting back to take fastest time on two of the 15 stages. Fourth place kept him in contention, although Subaru's Burns was now nine points ahead in the title chase. Uwe Nittel won Group N in his Evo VI.

2000 Monte Carlo Rally (Mäkinen).

In Argentina, many of the top runners went out. However, the Mitsubishi works cars gave a reasonable showing, while the Group N machines continued to be invincible, winning the Class yet again and coming ninth and tenth overall.

In Greece, Mäkinen lost a rear wheel, like many of his fellow drivers, including Loix, who went on stage one; those that didn't lose a rear wheel seemed to lose a front one in a crazy war of attrition. At least Mitsubishi continued its unbeaten run in Group N, thanks to Gabriel Pozzo. Matters didn't improve in NZ: Mäkinen crashed into a wall and decided to withdraw, as the new differential was not working well; Loix also retired. Lancers finished seventh, eighth, ninth and tenth, however, with Group N again going to Mitsubishi thanks to Manfred Stohl.

On the 1000 Lakes, the works cars were prepared to Evo VI½ specification, easily distinguished by the new front air dam. Loix had a minor accident on the first stage of the second day, puncturing his radiator, which resulted in the engine overheating. Mäkinen complained about the longer suspension travel, winning only one stage en route to fourth place. Jani Paasonen, in a Ralliart Finland-entered car, kept up Mitsubishi's season-long unbeaten run in Group N. It was Paasonen who took the Class in Sweden a few months earlier, incidentally.

Mäkinen was back on form in Cyprus, but he couldn't make up for the time lost by transmission problems early on. Loix kept going, but finished over eight minutes down on the winning Ford of Carlos Sainz. Gustavo Trelles took Group N.

On the Tour de Corse, both cars crashed out – Loix on stage one, and Mäkinen on the last day (thus ending any realistic hopes of claiming another Championship), although Manfred Stohl took Group N honours. Loix gave a better performance in San Remo, but Mäkinen was off the pace. If nothing else, a locally-entered privateer upheld Mitsubishi's unbeaten record in Group N.

There was more bad luck in

The Lancer Evolution VI cockpit.

2000 Rally of Argentina (Loix).

2000 Safari Rally (Mäkinen).

2000 1000 Lakes Rally (Mäkinen).

the Ralliart camp in Australia. Mäkinen was excluded after leading in the final stages, judged to have an illegal turbocharger, even though it offered no advantage. It was a strange situation, but Grönholm, the man leading the title chase, was running second anyway, which would still have put the Championship beyond the Finn's reach. Loix had clutch failure, although Gustavo Trelles won Group N.

Loix crashed out of the RAC early on, rolling his car on stage two, but Mäkinen recovered after a rough start – pushing hard (too hard, sometimes!) was the only way to stay with the front runners. Manfred Stohl took Group N honours, giving Mitsubishi 14 wins from 14 rounds in the Class. An amazing record.

Tommi Mäkinen came fifth in the Drivers'

2000 Rally of New Zealand (Loix).

Cars for the 2000 San Remo Rally.

Championship after a disappointing season, at least by his standards. Loix trailed home 15th in the title chase, while Mitsubishi came fourth on 43 points – the gap between traditional Group A and the new WRCar machines was finally beginning to tell. Peugeot was declared champion on 111 points, followed by Ford (91) and Subaru (88).

2000 RAC Rally (Mäkinen).

2000 Cyprus Rally (Loix).

2000 RALLY RECORD

No	Driver/co-driver	Position	Reg No (Group)
Monte Carlo (20-22 January)			
1	Tommi Mäkinen/Risto Mannisenmäki	1st	V2 MMR (Gp A)
2	Freddy Loix/Sven Smeets	6th	V22 MMR (Gp A)
Sweden (10-13 February)			
1	Tommi Mäkinen/Risto Mannisenmäki	2nd	V4 MMR (Gp A)
2	Freddy Loix/Sven Smeets	8th	T2 MMR (Gp A)
Safari (25-27 February)			
2	Freddy Loix/Sven Smeets	dnf	S33 TMR (Gp A)
1	Tommi Mäkinen/Risto Mannisenmäki	dnf	S3 TMR (Gp A)

No	Driver/co-driver	Position	Reg No (Group)
Portugal (16-19 March)			
2	Freddy Loix/Sven Smeets	6th	T2 MMR (Gp A)
1	Tommi Mäkinen/Risto Mannisenmäki	dnf	V4 MMR (Gp A)
Spain (31 March-2 April)			
1	Tommi Mäkinen/Risto Mannisenmäki	4th	V2 MMR (Gp A)
2	Freddy Loix/Sven Smeets	8th	V22 MMR (Gp A)
Argentina (11-14 May)			
1	Tommi Mäkinen/Risto Mannisenmäki	3rd	W2 MMR (Gp A)
2	Freddy Loix/Sven Smeets	5th	S2 TMR (Gp A)
Acropolis (9-11 June)			
1	Tommi Mäkinen/Risto Mannisenmäki	dnf	V4 MMR (Gp A)
2	Freddy Loix/Sven Smeets	dnf	W4 MMR (Gp A)
New Zealand (13-16 July)			
1	Tommi Mäkinen/Risto Mannisenmäki	dnf	W2 MMR (Gp A)
2	Freddy Loix/Sven Smeets	dnf	S2 TMR (Gp A)
1000 Lakes (18-20 August)			
1	Tommi Mäkinen/Risto Mannisenmäki	4th	W9 MMR (Gp A)
2	Freddy Loix/Sven Smeets	dnf	V22 MMR (Gp A)
Cyprus (8-10 September)			
1	Tommi Mäkinen/Risto Mannisenmäki	5th	V4 MMR (Gp A)
2	Freddy Loix/Sven Smeets	8th	W4 MMR (Gp A)
Tour de Corse (29 September-1 October)			
1	Tommi Mäkinen/Risto Mannisenmäki	dnf	X3 MMR (Gp A)
2	Freddy Loix/Sven Smeets	dnf	X33 MMR (Gp A)
San Remo (20-22 October)			
1	Tommi Mäkinen/Risto Mannisenmäki	3rd	W9 MMR (Gp A)
2	Freddy Loix/Sven Smeets	8th	V4 MMR (Gp A)
Australia (9-12 November)			
2	Freddy Loix/Sven Smeets	dnf	S2 TMR (Gp A)
1	Tommi Mäkinen/Risto Mannisenmäki	exc	W2 MMR (Gp A)
RAC (23-26 November)			
1	Tommi Mäkinen/Risto Mannisenmäki	3rd	V22 MMR (Gp A)
2	Freddy Loix/Sven Smeets	dnf	V4 MMR (Gp A)

The 2001 rally season

The sixth generation Lancer – the Lancer Cedia – had been launched in May 2000, coming with a longer wheelbase, and a very strong four-door sedan bodyshell. Front-wheel drive was the norm at this stage, but the inclusion of a transmission tunnel gave a definite clue of what was to come. MacPherson struts were used in the front suspension, while a multi-link set-up was adopted at the rear.

Mitsubishi duly announced the Lancer Evolution VII at the end of January 2001. Carrying the CT9A chassis code, the conventional four-door saloon styling theme was continued, but the new body was 175.4in (4455mm) long, 69.7in (1770mm) wide, and had an overall height of 57.1in (1450mm); ground clearance was 5.5in (140mm). The wheelbase was stretched to 103.3in (2625mm), while the track was 59.6in (1515mm) front and rear (although it was listed at 59.0in, or 1500mm, on the RS). As for the weight, the GSR tipped the scales at 3080lb (1400kg), with the RS being 176lb (80kg) less – the gap between the two models was therefore less than before. The fuel tank was a fraction smaller, with a capacity of 10.6 imperial gallons, which equates to 48 litres.

Only the front doors, roof and bootlid were carried over directly from the Lancer Cedia, for even though the main section of the floorpan was the same, the rear part was new. Other interesting features on the body worthy of note included the single blade on the rear spoiler, which not only had a greater surface area, but was also lighter, and the way in which the foglamps were built into the headlights to allow extra room for cooling apertures in the integrated bumper/air dam. It should also be noted that the door glass was some 10 per cent thinner to save weight.

The engine was still rated at 280bhp which is Japan's voluntary limit for domestic vehicles. Contemporary reports stated that there was strong torque between 3000 and 7000rpm; 80 per cent of maximum torque was available from just 2000rpm.

Both the GSR and RS received new turbochargers with smaller nozzles: the GSR had the TD05HR-16G6-9.8T with Inconel blades, while the RS had titanium-aluminium alloy turbines in its TD05HRA-16G6-9.8T unit, although this high response turbocharger could be bought as a 50,000 yen option for the GSR. In addition, both the oil cooler and intercooler capacities were increased.

The Evolution VII employed the W5M51 transmission, which had a lighter shift action despite a pull-type clutch. The GSR's gear ratios were 2.928 on first, 1.950 on second, 1.407 on third, 1.031 on fourth, and 0.720 on fifth. The RS came with a 2.785 first gear, but all the other ratios were the same on the standard gearbox. However, the RS also had the option of different third, fourth and fifth cogs, listed at 1.444, 1.096 and 0.825 respectively. All cars came with a 4.529:1 final-drive ratio.

The press release – and those issued in the past – mentioned brake dimensions that are frankly impossible. The outside diameters were certainly impressive, but actually 12.6in (320mm) up front, and 11.8in (300mm) at the back. Smaller brakes, wheels and tyres came as standard on the RS, with 205/65 rubber on 15-inch rims – a marked difference to the GSR's 8J x 17 alloys shod with Yokohama Advan tyres. Cars with this latter combination (available as an option on the RS) had an additional front crossmember support.

The Evo WRC. With a bore and stroke of 85.5 x 86.9mm, the turbocharged 1996cc engine developed 300bhp at 5500rpm and 397lbft of torque at 3500rpm. Power was taken through a six-speed gearbox, and to keep it all in check there were massive ventilated discs at each corner (eight-pot calipers were fitted up front, with four-pot versions on the back). As with earlier works cars, Lasse Lampi was involved in the testing of the Lancer Evolution WRC; Tommi Mäkinen rarely did any testing – apparently, he just drove whatever he was given without complaint!

Anti-roll bars were fitted at both ends (0.94in in diameter up front and 0.87in at the back, or 24mm and 22mm, respectively), and there were a lot of aluminium components in the rear suspension. The front lower arm was also produced from this material, thus reducing the vehicle's unsprung weight.

The ACD system continuously monitored the speed of each wheel, longitudinal and lateral G, and steering and throttle positions to determine how much the centre differential needed to be locked, if at all. It ensured that the Evo VII's cornering posture remained stable throughout a turn, whilst improving traction coming out of one. The three settings, selected via a dash-mounted switch, helped the driver to fine-tune the system.

Meanwhile, the cars that contested the 2001 Monte Carlo Rally had 295bhp and almost 400lbft of torque under the bonnet. To keep this power in check, the front brakes were uprated with eight-pot calipers (previously six-pot), although the four-pot versions were retained for the rear.

The driver line-up stayed the same and, as in the previous year, Loix was running a 'Carisma GT' for the first part of the season. The 2815lb (1280kg) Evo 'VI½' won in Monte Carlo, Portugal and Africa, but it was now ostensibly outclassed. Therefore, biting the bullet and moving from the Group A arena to follow the newer World Rally Car rules

started to make sense, but the latest Lancer was the Cedia, which was substantially bigger and heavier than the fifth generation model, so perhaps not an ideal starting point for a competition car. There would be a lot of sleepless nights for the Mitsubishi engineers ...

However, at the end of July, Mitsubishi Ralliart issued the following press release:

"Mitsubishi Motors is embarking on a new and exciting chapter in its motorsport history by creating its first World Rally Car. Known as the Mitsubishi Lancer Evolution WRC, it will make its first FIA World Rally Championship appearance on Italy's San Remo Rally. It is designed to be the most advanced, most sophisticated competition car that Mitsubishi has yet produced.

"Switching to the World Rally Car regulations from the established Group A category represents a major change of policy for Mitsubishi Motors. The Group A rules allow significantly fewer modifications to the standard road car,

2001 Swedish Rally (Rådström).

a characteristic that Mitsubishi has welcomed, because it has cherished the direct link between the cars it sells in the showroom and the cars with which it competes around the world long after all other manufacturers decided to produce highly specialised cars purely for rallying.

"The team feels obliged to adopt a new approach, as Mitsubishi Ralliart Chief Engineer, Bernard Lindauer, explained: 'Now we have reached the point where the imbalance between Group A and World Rally Car regulations, and the resulting increased performance of World Rally Cars, puts Mitsubishi in the situation where we could not have all the development facilities available. The only way for us to improve performance and to gain more post-development benefit on the standard car was to follow World Rally Car rules.'

"The Mitsubishi Lancer WRC is closely based on the Lancer Cedia four-door saloon in the Japanese market, but it can exploit the extra freedom offered by the World Rally Car rules in a number of key areas. Mitsubishi Ralliart engineers have been able to change the suspension, some

2001 Monte Carlo Rally (Mäkinen).

elements of the bodywork and the weight distribution. It will, however, retain many components from the existing World Championship-leading Lancer Evolution, and it is therefore an exciting blend of new and well-proven technology.

"The Lancer Evolution WRC was tested for the first time on 24 July, by Marlboro Mitsubishi Ralliart's Belgian driver, Freddy Loix, at the Millbrook Proving Ground north of London. The test went according to plan, the car showing immediate promise.

"Loix commented: 'It's exciting to be involved in the World Rally Car project, and it was great to drive the car for the first time. The regulations allow us greater flexibility in certain areas and I think this, combined with elements of our current car, means we can look forward to a really competitive package ... The engineers have already made good steps with the engine and wheel travel at the front, and the weight distribution is much better. It feels different, not hugely, but still different.'

"Test work will concentrate on asphalt in the early stages, as the car is being introduced to the World Championship in Italy. Ralliart plans to carry out between

2001 Cyprus Rally (Loix).

2001 Rally of Portugal (Mäkinen).

2001 Acropolis Rally (Mäkinen).

2001 1000 Lakes Rally (Loix).

3000 to 4000 kilometres [approximately 1800 to 2400 miles] of testing as it prepares for a new venture in Mitsubishi motorsports history.

"After the first run, Lindauer added: 'Straight out of the box, the new car worked well, but we must be realistic that we have a lot of work ahead of us. Between now and San Remo there are five tests scheduled [both on asphalt and gravel]. With World Rally Car regulations, the weight distribution and increased suspension travel means the car should be a lot better, but there is always a question

of reliability with new cars. However, we have always had reliable parts in the past and done simulation tests in conjunction with the Japanese engineers, so hopefully any problems will be kept to a minimum.'

"After its debut in San Remo, both Tommi Mäkinen and Freddy Loix will contest the remaining three rounds in the Lancer Evolution WRC."

For Mitsubishi, there were three key areas in the new car's development. The World Rally Car regulations

2001 Rally of New Zealand (Mäkinen).

allowed the powerplant to be moved up to one inch (25mm) from its usual location, thus allowing weight distribution to be optimised. Secondly, greater suspension travel allowed the engineers to build in greater stability and traction, especially on rough roads – the longer wheelbase of the Lancer Cedia would also aid stability, of course. And thirdly, more modifications could be applied to the engine.

Ultimately, the power unit was moved back as far as regulations would allow, improving both handling and tyre wear, while the suspension was given more travel. In fact,

the rear end was completely new, with MacPherson struts being used, like those already fitted to the front, while the wheelarches were enlarged at both ends. As for the engine, parts were lightened, internal friction was reduced, and the inlet and exhaust manifolds were modified to give better throttle response. In addition, the intercooler was relocated to increase its efficiency, thus giving a subtle boost in power and torque output.

The Lancer Evolution WRC was displayed at the Grimaldi Forum in Monaco just before the San Remo Rally.

2001 Australian Rally (Mäkinen).

Mäkinen went on record saying: "We are realistic that our car is new and still unproven in competition, but it felt good straight away when I tested it." Homologated on 1 October 2001, only time would tell ...

Looking back on the 2001 season in detail, Mäkinen led the Monte from the end of the second day to take a convincing victory, although it is interesting to note that Loix claimed the fastest time on one of the stages. A gaggle of Group N Evo VI drivers took from ninth place onwards, giving Mitsubishi Class honours once again.

Two detours into the Swedish countryside kept Loix trailing the leaders, and Mäkinen went off the road near the end whilst lying second. Thomas Rådström saved Mitsubishi's honour, though, with a fine second place in a 'Carisma GT.' To nobody's surprise, Group N fell to a Lancer driver.

In a muddy Portugal, Mäkinen led in all but three of the stages to claim the victor's spoils – both the Finn and Mitsubishi were in the number one slot of the WRC league tables, despite Loix going out with clutch trouble (he had been lying fifth). The Evo drivers complained about a lack of grip in Spain. However, Mäkinen still came third, with his Belgian team-mate fourth. Mäkinen wasn't really on form in Argentina, so was happy with his fourth place. After the event, Ford closed the gap on Mitsubishi to within just eight points. Carlos Sainz, who came third, was also catching up, finishing the event with 22 points against Mäkinen's 27. Lancer drivers continued to dominate Group N, holding the top six places in the league table (Gabriel Pozzo won the South American round, as he had in Spain).

In Cyprus, Mäkinen went off the road on the first day; luckily, trees stopped the car going too far down the hillside. Despite various problems, Loix survived to take fifth, while Gustavo Trelles took Group N in his Evo VI. Mäkinen was now ahead in the Championship by a single point, while Ford moved into the top spot, thanks to Colin McRae's win.

Loix faced problems again in Greece, including gearbox troubles and a damaged intercooler, but limped on to a top-ten finish. Mäkinen finished a distant fourth, meaning he now shared the lead with McRae in the WRC title chase. Pozzo won Group N (13th overall).

The Safari was good for Mitsubishi this year. Not only did Mäkinen come first (he led from the second stage), but McRae retired. It was a fine display from the Flying Finn, beating the second-placed Peugeot by no less than 12 minutes! Loix had engine trouble, but a cylinder head change kept him in the running; given the circumstances, his fifth place was quite an achievement.

2001 San Remo Rally (Mäkinen).

The win on the Safari put Mäkinen ten points clear of the Scot (his nearest competitor), with Mitsubishi six points ahead of Ford in the Manufacturers' title hunt. Group N cars dominated the category, with sixth, seventh, ninth and tenth overall. Thus, in Group N, Mitsubishi drivers had taken the chequered flag on all of the first eight events.

Mäkinen's home event turned out to be very disappointing. He crashed on the first stage, sustaining damage that led to his retirement. Loix could do no better than tenth, although Group N fell to Mitsubishi.

New Zealand witnessed a sad end to the Evo VI's career – both works drivers, and Toni Gardemeister in a third Ralliart-prepared car (V2 MMR), were never on the pace, although Manfred Stohl took Group N. Mäkinen and McRae were both on 40 points at the end of the event, with Ford ten ahead.

The San Remo Rally marked the debut of the Lancer Evolution WRC. Mäkinen lost a wheel on the final day, though he was only 11th before the crash and never really in with a shout. However, while Ford pulled further away, Colin's poor performance left the points standing unchanged in the Drivers' Championship.

On the Tour de Corse, Mäkinen crashed out heavily, injuring his co-driver. Two punctures cost Loix a lot of time in the early stages; in fact, he was lying 73rd at one point! Mäkinen was at least lucky that McRae finished out of the points, so they were still joint leaders. Trelles won the Group N category.

Tommi was sixth in Australia, but McRae was fifth, allowing the Scot to pull one point ahead; Burns had also caught up by now, just one point behind Mäkinen. Loix had electrical problems, which delayed his progress. Thus, in the Manufacturers' battle, Mitsubishi's hopes of winning were now completely out the window, although the Japanese firm remained unbeaten in Group N.

The final event of the season, the Rally of Great Britain, saw four drivers still in contention for the WRC Drivers' title: Tommi Mäkinen, Colin McRae (Ford), Richard Burns (Subaru), and Carlos Sainz (Ford). There was another battle, too: Mäkinen, Sainz, McRae and Juha Kankkunen had each won 23 events in the past; the first to win 24 would become the most successful driver in WRC history.

Mäkinen, partnered by Finland's Kaj Lindström for the event, was typically candid when asked about his chances: "I don't know why we haven't had that many good results

The rival teams – Ford

Ford had always been involved in rallying, directly, and then indirectly following its win in the 1979 World Championship. The Ford Escort RS was a regular sight, though, before giving way to the Group B RS200 in 1986. The Group B project was ill-timed, of course, and 4WD Sierras became Ford's weapon of choice in Group A until the works team went back to the Escort RS for 1993. This duly evolved into the Escort WRC, although it was replaced by the Focus WRC for the 1999 season. This was a much more competitive machine from the off, and after allowing Ford to come second in 2000, 2001, 2002 and 2004, it would duly deliver the UK-based team the Manufacturers' title in 2006 and 2007.

2001 RAC Rally (Mäkinen).

in Britain. I like the stages and I think we can go well, especially as our new Lancer WRC is getting better all the time. Still, the British drivers will not be easy to beat."

2001 Safari Rally (Mäkinen).

Freddy Loix's Lancer went out with transmission trouble halfway through the rally (he had been in 11th place), while Mäkinen crashed out almost immediately. Sadly, the November event was to be Mäkinen's last for the Ralliart team. Not only that, Subaru won Group N to deny Mitsubishi a clean sweep. Ultimately, the top Lancer was driven by Ramón Ferreyros, who finished 12th in a privately-entered Group A machine.

Burns' third place on the RAC gave him the World Championship, despite winning only one event during the season. He finished on 44 points, with McRae getting 42 from three wins, and Mäkinen scoring 41, also after three victories.

It may have been a different story if Mitsubishi had used the older but well-proven cars, in view of how close the title fight was. In all honesty, I thought it was a strange decision to use a completely new model so late in the WRC calendar, but have since learnt that there was little choice. The decision ultimately stemmed from an agreement made with the FIA in August 1999 that the Japanese company would enter the World Rally Car fold as soon as possible if the FIA would allow it to continue with the traditional vehicles. As such, the debut of the Evolution VII was long overdue, and couldn't have waited until the 2002 season.

In the Manufacturers' title chase, Peugeot won in fine style, followed home by Ford, then Mitsubishi and Subaru. In Group N, however, Mitsubishi was unbeatable, with all of the top finishers driving Lancers; Gabriel Pozzo taking the Class with ease.

2001 RALLY RECORD

No	Driver/co-driver	Position	Reg No (Group)
Monte Carlo (19-21 January)			
7	Tommi Mäkinen/Risto Mannisenmäki	1st	X6 MMR (Gp A)
8	Freddy Loix/Sven Smeets	6th	X3 MMR (Gp A)
Sweden (9-11 February)			
19	Thomas Rådström/Tina Thorner	2nd	S44 TMR (Gp A)
8	Freddy Loix/Sven Smeets	13th	X3 MMR (Gp A)
7	Tommi Mäkinen/Risto Mannisenmäki	dnf	W9 MMR (Gp A)
Portugal (8-11 March)			
7	Tommi Mäkinen/Risto Mannisenmäki	1st	X66 MMR (Gp A)
8	Freddy Loix/Sven Smeets	dnf	V22 MMR (Gp A)
Spain (23-25 March)			
7	Tommi Mäkinen/Risto Mannisenmäki	3rd	X66 MMR (Gp A)
8	Freddy Loix/Sven Smeets	4th	X3 MMR (Gp A)
Argentina (3-6 May)			
7	Tommi Mäkinen/Risto Mannisenmäki	4th	X2 MMR (Gp A)
8	Freddy Loix/Sven Smeets	6th	W9 MMR (Gp A)
Cyprus (1-3 June)			
8	Freddy Loix/Sven Smeets	5th	V22 MMR (Gp A)
7	Tommi Mäkinen/Risto Mannisenmäki	dnf	X66 MMR (Gp A)
Acropolis (15-17 June)			
7	Tommi Mäkinen/Risto Mannisenmäki	4th	W2 MMR (Gp A)
8	Freddy Loix/Sven Smeets	9th	V2 MMR (Gp A)
Safari (19-22 July)			
7	Tommi Mäkinen/Risto Mannisenmäki	1st	W9 MMR (Gp A)
8	Freddy Loix/Sven Smeets	5th	S33 TMR (Gp A)
1000 Lakes (24-26 August)			
8	Freddy Loix/Sven Smeets	10th	X2 MMR (Gp A)
7	Tommi Mäkinen/Risto Mannisenmäki	dnf	X6 MMR (Gp A)
New Zealand (21-23 September)			
7	Tommi Mäkinen/Risto Mannisenmäki	8th	X66 MMR (Gp A)
8	Freddy Loix/Sven Smeets	11th	V22 MMR (Gp A)

No	Driver/co-driver	Position	Reg No (Group)
San Remo (5-7 October)			
8	Freddy Loix/Sven Smeets	12th	Y4 MMR (Gp A)
7	Tommi Mäkinen/Risto Mannisenmäki	dnf	Y5 MMR (Gp A)
Tour de Corse (19-21 October)			
8	Freddy Loix/Sven Smeets	12th	Y4 MMR (Gp A)
7	Tommi Mäkinen/Risto Mannisenmäki	dnf	Y5 MMR (Gp A)
Australia (2-4 November)			
7	Tommi Mäkinen/Timo Hantunen	6th	Y7 MMR (Gp A)
8	Freddy Loix/Sven Smeets	11th	Y4 MMR (Gp A)
RAC (22-25 November)			
8	Freddy Loix/Sven Smeets	dnf	Y4 MMR (Gp A)
7	Tommi Mäkinen/Kaj Lindström	dnf	Y9 MMR (Gp A)

The 2002 rally season

The release of an automatic Evo VIII (the GT-A) in the early part of 2002 was an interesting move, although it had no bearing on the works rally programme at the time. With Tommi Mäkinen leaving for Subaru, and Freddy Loix also leaving the camp, it was the driver line-up that was the biggest news coming from the Ralliart offices.

The new number one driver was François Delecour. Partnered by Daniel Grataloup in the passenger seat, the flamboyant Frenchman had been with Ford the previous season, and had won a number of WRC events for the marque in the past. He celebrated his 40th birthday during his first season with Mitsubishi. "It will be very good to drive for Mitsubishi, a team that has produced a four-times World Champion," commented Delecour. "It's a fantastic opportunity for me. I realise there is a new car, but with Mitsubishi's history, I'm positive it will be successful."

The second new member was Alister McRae. The author grew up watching his father Jimmy in action, and his older brother Colin will be familiar to those who have followed the rally scene in more recent years. Born in Scotland in 1970, he started his rally career in 1988, and had recently been with Hyundai (driving the Accent WRC) partnered, as always, by David Senior.

The 2002 Monte Carlo Rally was full of drama, as usual. McRae hit a wall on the last night, but at least managed to limp home to finish the event in 14th place. Delecour declared himself happy with his new mount, despite finishing outside the points – his sights were firmly set on a good performance in Corsica. Meanwhile, the winning Citroën was demoted to second place, handing Mäkinen another Monte victory.

François Delecour.

Alister McRae.

In Sweden, Jani Paasonen had been going well in a third Evolution WRC but lost a wheel after hitting a rock hidden by thick snow; he finished in 14th position. Delecour went off the road into a ditch, thus ending his chances of a decent finish on the first leg, although he continued nonetheless to gain more experience with the Lancer. McRae was eighth at the end of the second day, but moved up to take fifth at the expense of his brother, who was only a fraction behind. McRae and Mitsubishi therefore got into the points, scoring two and three respectively.

A new time slot for the Tour de Corse meant unknown conditions for all of the teams. Delecour said he felt the car was much quicker. He finished the first day in 11th place, although had been running higher until rain hit the event. McRae also fell foul of the weather, ending the first leg in 18th. The Mitsubishi pair were in and around the top ten again on the second day, with Delecour finishing the leg three places better off (although nearly four minutes down on the leading Peugeot), while McRae ended the day in 12th – his charge being delayed by a broken damper on Stage 12. The final day was much the same, with few major retirements until the end. Delecour ultimately finished seventh, thus failing to score for himself, but taking Mitsubishi's tally to five points (fourth place in the Manufacturers' title chase, although a long way behind Peugeot). McRae came home in tenth place.

French cars dominated the first day in Spain; Delecour was trailing the leader by three-and-a-half minutes, while punctures delayed McRae (he ended the leg five minutes down). Nothing much changed on day two, as many of the

2002 Monte Carlo Rally (Delecour).

stages were cancelled due to spectator safety concerns. On the final leg, Delecour moved up to ninth, but McRae spun on one stage, keeping him back down the field. At least Delecour's efforts gave Mitsubishi another point.

The Japanese team ran a third car for Jani Paasonen in Cyprus (with McRae and Delecour suffering with brake problems and broken driveshafts in the early stages (conditions were extremely rough around Mount Olympus), it was Paasonen who led the Mitsubishi challenge until early on the second day when he retired with broken steering. McRae rolled but continued until SS11 when he lost drive. The Lancers were never really on the pace, although Delecour pulled a good time out of the bag on stage 11 and finished the event in 13th place.

recorded some consistent times to finish day one in 11th, one place ahead of Delecour. A fire at the team's hotel in the night couldn't have helped the Lancer drivers; McRae dropped two places, but at least the French pairing held on to 12th. Unfortunately, McRae went out at the start of day three with broken steering (he cut a corner and paid a high price), although Delecour had a clean run to take 11th.

In reality, Mitsubishi was pinning its hopes on the 'Version 2' car, using the early rallies of 2002 for testing and the collection of data, but there were still hopes that the Ralliart team could come away from the Safari with a decent reward for its efforts. Sadly, Delecour's engine overheated after the cooling fan packed up. McRae was fast on some stages, but dogged by a series of problems (including a shattered brake disc): he had to settle for 14th at the end of the first day. The Scot continued to charge, however, moving up to ninth going into the third day – the position in which he finished, claiming points in his African debut. Only 12 cars survived the event, won this year by Alister's brother, Colin.

Mitsubishi's new drivers, in what was still a new car, had made slow progress during the first part of the 2002 season. The other leading teams had quite a head start on Mitsubishi when it came to making a World Rally Car go faster for longer. The fact that the Evo VII-based machine had, in the main, been reliable, at least provided a glimmer of hope that more trophies would be added to the Ralliart collection.

On the day before the 1000 Lakes Rally started, the

With stronger driveshafts and revised engine mapping to suit the high altitudes of the Rally Argentina, Mitsubishi expectations were high. Indeed, Delecour was going well until a broken suspension forced his retirement on SS7, and McRae would have done better had he not been hampered by tyre and steering problems. To add to the Scot's misery, he lost turbo boost at the start of the second day, dropping him further down the field. However, there was a great deal of drama on the final day – Mäkinen had a big accident, dashing his hopes of victory, Grönholm was excluded (robbing him of victory), and then, at post-event scrutineering, Burns' car was deemed not to conform with homologation papers. Peugeot didn't appeal against the decision, meaning Sainz won, and McRae moved up to eighth; the Group N Mitsubishi of Ramón Ferreyros came tenth.

As the WRC circus moved to Greece, Alister McRae

2002 Catalunya Rally (Delecour).

Lancer Evolution WRC2 made its debut in Savutuvan Apaja, Finland. The latest car was lighter, and had better weight distribution and a lower centre of gravity. It also had improved cooling and aerodynamics thanks to a series of bodywork refinements, and a revised suspension package that offered longer travel and enhanced rigidity. The active 4WD system was also updated, while the 4G63 engine received a number of modifications to reduce weight (the crankshaft, flywheel and other internal parts were lightened) and improve low- to mid-range response, the latter via the adoption of a new exhaust manifold and single-scroll

turbocharger. The 1996cc unit gave 300bhp at 5500rpm, and 397lbft of torque at 3500rpm, delivered to all four wheels through a six-speed INVECS-type transmission. The Enkei alloys were shod with Michelin rubber, hiding a six-pot caliper braking system up front, and a four-pot one at the rear.

The extensive testing in France and Finland seemed to be paying off, with Jani Paasonen making the most of his local knowledge in a third works car. Delecour was sidelined early in the event after a suspension breakage following a jump, but Paasonen and McRae stayed in touch

95

2002 Acropolis Rally (Delecour).

to finish day one in 13th and 17th respectively. A fire halted the Scot's charge, however, and he ultimately retired on the final day with suspension trouble. At least Paasonen was having a good run, being fourth fastest on the first stage of day three, helping him to move into eighth place at the end of the event.

Germany provided the new car with a very different kind of test. This was the first Rallye Deutschland to be included as a WRC qualifying round, but it was not kind to the Ralliart team. McRae retired early on when the engine died, and Delecour was unable to produce the sort of pace expected from him until the second day, when he was second fastest on one of the stages. Unfortunately, Delecour was down on power for the final day, dropping him down the field again after a last minute push from Ford's Carlos Sainz.

Nevertheless, Ralliart's Chief Engineer, Bernard Lindauer, was confident going into San Remo. Indeed, Delecour was running tenth until a lack of turbo boost slowed him down. McRae was struggling, and before the start of the second day, Team Manager Derek Dauncey withdrew the Scot – apparently he'd injured himself on a mountain bike prior to the event. Delecour picked up the pace again to finish tenth overall, scoring a point for Mitsubishi along the way. This moved the Japanese manufacturer into fourth in the title chase, but it was still a long way behind Peugeot, Ford and Subaru.

Paasonen was drafted in to replace McRae for New Zealand, and this proved to be a good move. The Flying Finn clocked the fastest time on one stage (a first for both him and the new car), and was running third for part of the first day. Sadly, Paasonen left the road on day two,

2002 Safari Rally (McRae).

damaging the car and ending his rally in the process, and the best Delecour could manage was ninth position. Delecour had a rather nasty accident in Australia, and Paasonen was off the pace. He rolled the car on day two but continued to take ninth at the end of the event.

In the UK, Delecour had a new co-driver after Grataloup was injured in Oz, and Paasonen (again taking the place of McRae) tackled the RAC for the first time. Justin Dale got his first works drive in a third Lancer prepared for the event. Interestingly, fourth place in the WRC title chase was shared by Mitsubishi, Skoda and Hyundai going into the rally, so a good result was crucial. It was not to be, though, and after Dale crashed out early, the other two Ralliart entries dropped out on day two.

2002 was ultimately disappointing for the Mitsubishi team, as Hyundai scored a point on the RAC to secure fourth place in the Manufacturers' championship. Alister McRae was the only Ralliart driver to score a point, coming joint-14th at the end of the season, alongside Bruno Thiry and Juha Kankkunen on two points.

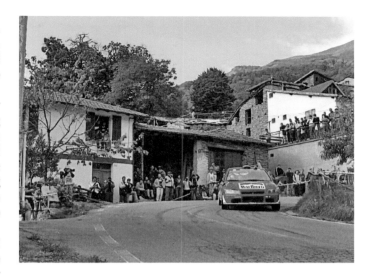

2002 San Remo Rally (Delecour).

2002 Australian Rally (Paasonen).

Left: 2002 1000 Lakes Rally (McRae).

No	Driver/co-driver	Position	Reg No (Group)
Monte Carlo (17-20 January)			
7	François Delecour/Daniel Grataloup	9th	KP51 RWO (Gp A)
8	Alister McRae/David Senior	14th	KP51 RWL (Gp A)
Sweden (1-3 February)			
8	Alister McRae/David Senior	5th	KP51 RWN (Gp A)
9	Jani Passonen/Arto Kapanen	14th	Y4 MMR (Gp A)
7	François Delecour/Daniel Grataloup	34th	Y9 MMR (Gp A)
Tour de Corse (8-10 March)			
7	François Delecour/Daniel Grataloup	7th	KP51 RWO (Gp A)
8	Alister McRae/David Senior	10th	KS51 XPN (Gp A)
Spain (21-24 March)			
7	François Delecour/Daniel Grataloup	9th	KP51 RWO (Gp A)
8	Alister McRae/David Senior	13th	KS51 XPN (Gp A)
Cyprus (19-21 April)			
7	François Delecour/Daniel Grataloup	13th	KU02 HJZ (Gp A)
8	Alister McRae/David Senior	dnf	KP51 RWN (Gp A)
9	Jani Passonen/Arto Kapanen	dnf	Y4 MMR (Gp A)
Argentina (16-19 May)			
8	Alister McRae/David Senior	8th	KS51 XPN (Gp A)
7	François Delecour/Daniel Grataloup	dnf	KP51 RWO (Gp A)
Acropolis (13-16 June)			
7	François Delecour/Daniel Grataloup	11th	KU02 HJZ (Gp A)
8	Alister McRae/David Senior	dnf	KP51 RWN (Gp A)
Safari (12-14 July)			
8	Alister McRae/David Senior	9th	KS51 XPN (Gp A)
7	François Delecour/Daniel Grataloup	dnf	KP51 RWO (Gp A)
1000 Lakes (8-11 August)			
9	Jani Paasonen/Arto Kapanen	8th	Y7 MMR (Gp A)
8	Alister McRae/David Senior	dnf	KR02 DLZ (Gp A)
7	François Delecour/Daniel Grataloup	dnf	KR02 DME (Gp A)

No	Driver/co-driver	Position	Reg No (Group)
Germany (22-25 August)			
7	François Delecour/Daniel Grataloup	9th	KR02 DLY (Gp A)
8	Alister McRae/David Senior	dnf	KR02 DLX (Gp A)
San Remo (19-22 September)			
7	François Delecour/Daniel Grataloup	10th	KR02 DLY (Gp A)
8	Alister McRae/David Senior	dnf	KR02 DLX (Gp A)
New Zealand (3-6 October)			
7	François Delecour/Daniel Grataloup	9th	KR02 DME (Gp A)
9	Jani Paasonen/Arto Kapanen	dnf	KR02 DLZ (Gp A)
Australia (31 October-3 November)			
9	Jani Paasonen/Arto Kapanen	9th	KP51 RWN (Gp A)
7	François Delecour/Daniel Grataloup	dnf	KR02 DME (Gp A)
RAC (14-17 November)			
9	Jani Paasonen/Arto Kapanen	dnf	KR02 DLX (Gp A)
7	François Delecour/Dominique Savignoni	dnf	KR02 DLY (Gp A)
8	Justin Dale/Andrew Bargery	dnf	KU02 HJZ (Gp A)

The 2003 rally season

While a new Evo VIII had been released both in the States (a first for the model), as well as Japan and Europe, at the start of 2003, rally fans were to be disappointed for the coming season, although the formation of Mitsubishi Motors Motor Sports (or MMSP) in Trebur, Germany, confirmed a willingness to continue the WRC programme.

"Mitsubishi Motors Corporation (MMC) announced today that it will restructure its motorsports engineering activities over the next 12 months to prepare for a full-fledged return to the FIA World Rally Championship in 2004. MMC will not take part in the WRC in 2003 to concentrate fully on a successful return to the event in 2004." This blunt press release, issued on 3 December 2002, was heartbreaking news, although it did go on to say that Ralliart Europe would enter a few rallies for development purposes.

This move came about from the realisation that Mitsubishi had relied on the old Group A car for too long, and been left behind in the WRCar arena. A year's break to regroup seemed the only way forward. Reporting to Mitsubishi stalwart Iwao Kimata, Sven Quandt was appointed Team Manager, with ex-Peugeot man Mario Fornaris as Chief Engineer, but – as the press release had stated – it was far from a full season of activity,

with the WRC2 having only a few outings and losing its Marlboro sponsorship along the way.

Alister McRae and Jani Paasonen got a couple of outings between them, but it was Kristian Sohlberg (born in Finland in 1978) who flew the flag for Mitsubishi on the most occasions. At least Mitsubishi was strongly represented in Group N, with 80 per cent of privateers choosing the LanEvo as their preferred mount.

The Swedish Rally kicked off the season for the Group A Lancers, with Kristian Sohlberg running a works-supported

2003 Rallye Deutschland (Sohlberg).

car. He drove a consistent rally, moving up to tenth place at one point, but finishing the event in 12th.

Sohlberg was in action again in New Zealand, but an accident five stages from the end put him on the retirements listing. Alister McRae was entered in a sister car, and finished in a creditable sixth place after a steady, calculated drive.

Germany offered Sohlberg the opportunity to show his talent in a completely different arena, but he was never really on the pace. Jani Paasonen did a much better job of staying with the leaders, but a heavy crash on SS11 ended his rally and left his co-driver badly injured.

The World Championship was obviously not an issue for Mitsubishi in 2003 (it was won by Citroën), and the Group N category was dominated by Subaru drivers. In the Asia-Pacific Championship, however, it was Evo VII driver Armin Kremer that clinched the title after a podium finish in the last round of the series.

The rival teams – Citroën

The Citroën team made its WRC debut in 1998 with a Xsara variant, although entries were few and far between until 2002, when more effort was made. This approach, slowly building up experience, allowed the Versailles-based equipe to take the World Championship crown in 2003, 2004 and 2005 with the Xsara WRC. After a short break, the works Citroën team returned to the WRC scene with the C4 in 2007, and the French outfit would ultimately take five more titles from 2008 onwards.

2003 RALLY RECORD

No	Driver/co-driver	Position	Reg No (Group)
Sweden (7-9 February)			
35	Kristian Sohlberg/Jakke Honkanen	12th	KR02 DLZ (Gp A)
New Zealand (10-12 April)			
32	Alister McRae/David Senior	6th	KP51 RWL (Gp A)
33	Kristian Sohlberg/Jakke Honkanen	dnf	KP51 RWN (Gp A)
Germany (25-27 July)			
34	Kristian Sohlberg/Jakke Honkanen	14th	KN52 XBC (Gp A)
32	Jani Paasonen/Arto Kapanen	dnf	KN52 XBB (Gp A)

The 2004 rally season

The Lancer Evolution VIII MR made its debut at the 2003 Tokyo Show, but the latest works rally machine was to be quite a different beast, with only a passing resemblance to the road cars carrying the flag.

The Lancer WRC04 was first displayed at the 2003 Essen Show in late-November. The new bodywork hid a completely new drivetrain, suspension and braking system – in fact, around 6000 parts were developed for the latest WRC challenger. The repositioned 1996cc engine, with its Garrett turbocharger, developed the same power and torque as the 'Version 2' car of 2002, but it was now transmitted to the 4WD system (with passive front, centre and rear differentials) via a Ricardo/MMSP five-speed sequential gearbox. Again, Enkei wheels were chosen to shroud the Brembo brakes, shod with Michelin rubber. In fact, all the WRC teams except Subaru ran with Michelin tyres (the SWRT preferred Pirellis), and all but Mitsubishi used OZ wheels.

Gilles Panizzi (born in Menton, France, in 1965) was the main driver, signed up in August 2003, with the second seat being shared between three young lions. Having made his WRC debut on the 1990 Monte Carlo Rally, and a rally winner on six occasions since, Panizzi brought with him

Gilles Panizzi (right) talking with Gianluigi Galli.

the experience needed to balance the fresh talent lined up alongside him.

Gianluigi Galli was born in Milan in 1973. Having made his rally debut in 1994, he was immediately on the pace in the Group N championship and the JWRC.

Spaniard Daniel Sola was two years younger than Galli. Sola started rallying in 1996, and took the JWRC crown in 2002.

Kristian Sohlberg drove the second works car on the kind of events that tend to suit Scandinavians. Having made his WRC debut in 2000, he was still very much a novice at this level, but he'd done enough in 2003 to convince new team boss Sven Quandt that he deserved the chance.

The season kicked off, as was tradition, with the Monte. At least the new car was on the pace, with Panizzi coming a solid sixth. Galli was fourth on one of the early stages, but went off the road and into retirement on the fifth.

Mitsubishi never really got into the groove in Sweden, with Panizzi retiring with mechanical problems on stage four, and Sohlberg following him back to the team bus with transmission maladies on the next stage. At least there was a Group N win courtesy of Mattias Ekström in an Evo VIII.

In Mexico, Galli went out quickly with suspension problems, while Panizzi stayed in eighth from day one. An amazing performance from Daniel Sola to take Group N honours (11th overall) was enough to earn him a works drive.

Both cars retired with electrical problems on the first stage of the New Zealand Rally, and in Cyprus, both cars were withdrawn midway through rally with engine trouble. However, Group N again fell to Mitsubishi, with Evo drivers leading the way in the PCWRC. Unfortunately, the works cars were trailing their rivals quite badly, already 52 points behind leaders Ford, and 28 behind their closest other maker, Peugeot.

In Greece, Sola dropped out after stage three, but Panizzi stayed in touch with the leaders and should have been in a strong points-scoring position until a mistake on

2004 Monte Carlo Rally (Panizzi).

the last day. At least there was another Group N victory for Mitsubishi.

Panizzi dropped out from the Rally of Turkey with electrical problems early on, leaving Galli to fly the Ralliart flag. The Italian had problems of his own, however, which forced him down the field before a good recovery on the final day.

Panizzi regularly put in good stage times in Argentina, but could only muster seventh place for all his efforts. Sohlberg retired towards the end of the second leg with gearbox trouble, which was a shame, as he'd managed to threaten those on the leaderboard by holding fifth place until then. Although there was no Group N trophy, Jani

2004 Monte Carlo Rally (Galli).

2004 Rally of Mexico (Panizzi).

Paasonen led the PCWRC title chase, with Mitsubishi stalwart Manfred Stohl in second, and Daniel Sola in third.

In Finland, Panizzi simply went through the motions, and was never really in a position to challenge the faster French cars, while Sohlberg finished the first day in ninth. He was improving his position, too, until an excursion led to his retirement. In Group N, however, there was a victory for occasional works driver Gigi Galli.

In Germany, Sola came third on the first stage but crashed out of the next. Panizzi also left the road, saying goodbye to the sixth place he'd held at the end of the first leg. Mitsubishi kept up its momentum in Group N, though, with a win courtesy of Xavier Pons and Oriol Julia; Paasonen was now well clear of his closest rival.

Then everything went wrong. The works cars stayed away from the first ever Rally Japan, giving Subaru the opportunity to claim a whitewash in both categories. The author was there in the press team with my old friend Peter Lyon, and can only say what a missed opportunity it was being absent from this home turf event.

There were still no works Lancers to be seen in Britain or Italy, although Galli was able to claim a Group N victory on the roads of Sardinia. Again, there were no works cars on the Tour de Corse, but LanEvo VII driver Xavier Pons (number 47) lifted the Group N trophy, putting him in touch with Paasonen at the head of the PCWRC table.

At least in Spain, there were three WRC04 machines, with Sola and Galli restoring Mitsubishi pride. Both drove a very solid event, with Panizzi bringing the third car home in a respectable 12th position.

Sadly, the works cars stayed away again for the final round of the season. Unfortunately, because of the way the championship worked, and despite Pons' second in Group N in Australia, the title went to a Subaru driver after Paasonen failed to deliver the goods. It was a disappointing end to a disappointing year.

The failure to complete all the rounds in the championship led to Mitsubishi not qualifying for a finishing position, although it's fair to say – barring a miracle – that

2004 Rally of New
Zealand (Sohlberg).

2004 Cyprus
Rally (Panizzi).

the team would have ended the season in fifth anyway, with only Peugeot being within any kind of realistic reach.

Gilles Panizzi ended the WRC title chase in 13th – not bad considering his lack of drives at the tail-end of the season. Gigi Galli came home in 15th, with Daniel Sola 24th, while Kristian Sohlberg failed to score any points. Jani Paasonen finished third in the PCWRC, with Xavier Pons fourth, and Manfred Stohl sixth.

2004 Rally of Argentina (Sohlberg).

106

2004 Rallye
Deutschland
(Panizzi).

2004 Rallye Deutschland (Sola).

Armin Kremer in Group N action.

No	Driver/co-driver	Position	Reg No (Group)
Monte Carlo (23-25 January)			
9	Gilles Panizzi/Hervé Panizzi	6th	KX53 BKY (Gp A)
10	Gianluigi Galli/Guido D'Amore	dnf	KX53 BKV (Gp A)
Sweden (6-8 February)			
10	Kristian Sohlberg/Kaj Lindström	dnf	KX53 BKO (Gp A)
9	Gilles Panizzi/Hervé Panizzi	dnf	KX53 BKU (Gp A)
Mexico (12-14 March)			
9	Gilles Panizzi/Hervé Panizzi	8th	KX53 BKY (Gp A)
10	Gianluigi Galli/Guido D'Amore	dnf	KX53 BKV (Gp A)
New Zealand (16-18 April)			
9	Gilles Panizzi/Hervé Panizzi	dnf	KR53 YPP (Gp A)
10	Kristian Sohlberg/Kaj Lindström	dnf	KX53 BKO (Gp A)
Cyprus (14-16 May)			
9	Gilles Panizzi/Hervé Panizzi	dnf	KX53 YPO (Gp A)
10	Kristian Sohlberg/Kaj Lindström	dnf	KX53 BKY (Gp A)
Acropolis (4-6 June)			
9	Gilles Panizzi/Hervé Panizzi	10th	KX53 YPP (Gp A)
10	Daniel Sola/Xavier Amigo Colon	dnf	KX53 BKO (Gp A)
Turkey (25-27 June)			
10	Gianluigi Galli/Guido D'Amore	10th	KX53 BKY (Gp A)
9	Gilles Panizzi/Hervé Panizzi	dnf	KR53 YPO (Gp A)
Argentina (16-18 July)			
9	Gilles Panizzi/Hervé Panizzi	7th	KN04 WLZ (Gp A)
10	Kristian Sohlberg/Kaj Lindström	dnf	KR53 YPP (Gp A)
1000 Lakes (6-8 August)			
9	Gilles Panizzi/Hervé Panizzi	11th	KR53 YPO (Gp A)
10	Kristian Sohlberg/Kaj Lindström	dnf	KX53 BKY (Gp A)
Germany (20-22 August)			
9	Gilles Panizzi/Hervé Panizzi	dnf	KN04 WMA (Gp A)
10	Daniel Sola/Xavier Amigo Colon	dnf	KX53 BKU (Gp A)

No	Driver/co-driver	Position	Reg No (Group)
Spain (29-31 October)			
10	Daniel Sola/Xavier Amigo Colon	6th	KR53 YPP (Gp A)
14	Gianluigi Galli/Guido D'Amore	7th	KN04 WLZ (Gp A)
9	Gilles Panizzi/Hervé Panizzi	12th	KR53 YPO (Gp A)

American advertising from Michelin
featuring Galli on the 2004 Monte.

The 2005 rally season

The Lancer Evolution IX was made available in early 2005 in three grades – the GSR, GT and RS. Once again, though, in order to be competitive, the works rally car was to be something rather different.

The Lancer WRC05 was a touch wider than its predecessor, with subtle revisions to the front and rear wings, and the rear quarter panel and bumper. The suspension, semi-automatic transmission and driveshafts were modified slightly, although the engine was much the same as that of the WRC04 (only the turbo wastegate and ECU were changed). Weighing in at 1250kg (2750lb), and with wider track measurements, the WRC05 now used Pirelli P-Zero tyres on Enkei rims following a successful series of tests conducted under the new Chief Engineer, Yasuo Tanaka.

Team boss Roger Estrada (reporting to the recently-appointed MMSP GmbH President, Isao Torii) saw to it that Harri Rovanperä was drafted in as the number one driver, lining up alongside Gilles Panizzi and Gigi Galli. Rovanperä was born in Finland in 1966, and had consistently finished in the top eight of the WRC series during the previous four seasons. A loose surface specialist, he was the perfect guy to team up with tarmac man Panizzi, and all-rounder Galli.

The Monte kicked off the season as usual and, despite problems with the new head restraint required by the FIA, Panizzi put in a blistering performance to give the new WRC05 model a podium position on its first outing. Rovanperä moved up

Harri Rovanperä.

2005 Monte Carlo Rally (Panizzi).

the leaderboard at a similar rate to finish the event in seventh.

Rovanperä drove a measured race in Sweden, but it was Galli that showed the early pace. Sadly, the Italian had to contend with a broken driveshaft and a faulty handbrake (not to mention a brief trip into the surrounding countryside), dropping him back down the field. He had been lying in third at one point.

Although the car was given a new active centre differential, a series of minor problems kept Mitsubishi from doing better in Mexico. A run of holed sumps, and an electrical problem for Rovanperä, saw to it that fifth was

2005 Rally of New Zealand (Galli).

2005 Swedish Rally (Rovanperä).

the best the team could come away with. At least an Evo VII took Group N honours.

In New Zealand, Rovanperä was running in a steady ninth until the final stage when both his rear tyres gave way, forcing him into retirement. Galli had one problem after another, but recovered well to claim eighth at the end of the event. Mitsubishi was still in with a chance of doing well in the WRC title chase at this point, lying fifth and only seven points down on second-placed Citroën.

Getting too close to the scenery put both cars out in Italy, and it was only thanks to the new Super Rally regulations that a finish was possible in Cyprus. Things were looking up in Turkey, however, with Galli actually leading the event (the first Mitsubishi driver to do so for four years!), but a turbo pipe broke loose and lost him a lot of time. Rovanperä hit a rock, rearranging his rear suspension; he restarted and ultimately finished tenth.

Galli had trouble with his turbo again on the Acropolis, but Rovanperä was undoubtedly the faster of the two works drivers on this tough event. Conditions didn't improve in Argentina, and although the Finn held on to take fifth,

Galli was sidelined by a suspension failure before gearbox trouble signalled the end.

Rovanperä was not able to shine on his home event, taking seventh after a superb opening stage, while Galli crashed out. This just about ended any hopes of Mitsubishi finishing in the top three of the WRC, as Peugeot, Citroën, Ford and Subaru pulled away.

Galli was in top form in Germany, finishing four minutes ahead of his team-mate, and only five down on Loeb's winning Citroën. Mitsubishi won the Group N category, although it was a Subaru driver that headed the PCWRC title chase at this stage in the proceedings.

The RAC was overshadowed by the death of Peugeot's Michael Park, but Rovanperä mastered the difficult conditions of the Welsh forests to take a good fourth place. The similar roads found on Rally Japan gave a similar result for the Finn (fifth position), and Galli was doing much better, holding fourth until an accident ended his event; Panizzi was never really on the pace in the third works car.

Three cars were entered again for the Tour de Corse,

2005 Rally of Italy (Rovanperä).

but, ironically, Galli (the top finisher) was not an allocated points scorer. Panizzi and Rovanperä failed to shine, the Frenchman ultimately retiring with electrical problems.

The Catalunya Rally was disappointing for Galli, as he hit a wall just after recording a fastest stage time. The car was stranded in a ditch, and that was that. Rovanperä never managed to threaten the leaders, although he put in a fantastic performance in Australia to take Mitsubishi's highest finish of the season; Galli also did well in the second car, claiming fifth, but it had all started to come together too late ...

The Mitsubishi team finished a gallant fifth, not so far behind Subaru, with Rovanperä coming seventh, Galli 11th, and Panizzi 15th. The Lancer Evolution IX was homologated as an FIA Group N machine in the autumn of 2005, adding heat to the long-running battle between Mitsubishi and Subaru in the PCWRC arena, but it was Impreza stalwart Toshi Arai that took the honours at the end of the season. The APC title went to a Mitsubishi man, though – Finland's Jussi Välimäki.

2005 Rallye Deutschland (Galli).

2005 Rally Japan (Rovanperä).

2005 Rally Japan (Galli).

2005 Tour de Corse (Panizzi).

2005 Acropolis Rally
(Rovanperä).

2005 Australian Rally
(Rovanperä).

2005 Australian Rally (Galli).

No	Driver/co-driver	Position	Reg No (Group)
Monte Carlo (21-23 January)			
10	Gilles Panizzi/Hervé Panizzi	3rd	KP54 GXY (Gp A)
9	Harri Rovanperä/Risto Pietiläinen	7th	KN04 WMD (Gp A)
Sweden (11-13 February)			
9	Harri Rovanperä/Risto Pietiläinen	4th	KN04 WLZ (Gp A)
10	Gianluigi Galli/Guido D'Amore	7th	KR53 YPO (Gp A)
Mexico (11-13 March)			
9	Harri Rovanperä/Risto Pietiläinen	5th	KN04 WMD (Gp A)
10	Gilles Panizzi/Hervé Panizzi	8th	KP54 GXY (Gp A)
New Zealand (8-10 April)			
10	Gianluigi Galli/Guido D'Amore	8th	KN04 WMC (Gp A)
9	Harri Rovanperä/Risto Pietiläinen	dnf	KN04 WLZ (Gp A)
Italy (29 April – 1 May)			
9	Harri Rovanperä/Risto Pietiläinen	dnf	KX05 AUR (Gp A)
10	Gianluigi Galli/Guido D'Amore	dnf	KR53 YPO (Gp A)
Cyprus (13-15 May)			
9	Harri Rovanperä/Risto Pietiläinen	7th	KN04 WMD (Gp A)
10	Gilles Panizzi/Hervé Panizzi	11th	KP54 GXY (Gp A)
Turkey (3-5 June)			
10	Gianluigi Galli/Guido D'Amore	8th	KN04 WMC (Gp A)
9	Harri Rovanperä/Risto Pietiläinen	10th	KN04 WLZ (Gp A)
Acropolis (24-26 June)			
9	Harri Rovanperä/Risto Pietiläinen	6th	KX05 AUR (Gp A)
10	Gianluigi Galli/Guido D'Amore	7th	KR53 YPO (Gp A)
Argentina (15-17 July)			
9	Harri Rovanperä/Risto Pietiläinen	5th	KN04 WMD (Gp A)
10	Gianluigi Galli/Guido D'Amore	dnf	KP54 GXY (Gp A)
1000 Lakes (5-7 August)			
9	Harri Rovanperä/Risto Pietiläinen	7th	KN04 WLZ (Gp A)
10	Gianluigi Galli/Guido D'Amore	dnf	KN04 WMC (Gp A)

(Continues overleaf)

Germany (26-28 August)

10	Gianluigi Galli/Guido D'Amore	5th	KR05 ZKJ (Gp A)
9	Harri Rovanperä/Risto Pietiläinen	10th	KR05 ZKL (Gp A)

RAC (16-18 September)

9	Harri Rovanperä/Risto Pietiläinen	4th	KN04 WMD (Gp A)
10	Gianluigi Galli/Guido D'Amore	13th	KP54 GXY (Gp A)

Japan (30 September-2 October)

9	Harri Rovanperä/Risto Pietiläinen	5th	KX05 AUR (Gp A)
10	Gilles Panizzi/Hervé Panizzi	11th	KR53 YPO (Gp A)
25	Gianluigi Galli/Guido D'Amore	dnf	KN04 WLZ (Gp A)

Tour de Corse (21-23 October)

18	Gianluigi Galli/Guido D'Amore	9th	KN04 WMC (Gp A)
9	Harri Rovanperä/Risto Pietiläinen	10th	KR05 ZKL (Gp A)
10	Gilles Panizzi/Hervé Panizzi	dnf	KR05 ZKJ (Gp A)

Spain (28-30 October)

9	Harri Rovanperä/Risto Pietiläinen	10th	KR05 ZKL (Gp A)
10	Gianluigi Galli/Guido D'Amore	dnf	KR05 ZKJ (Gp A)

Australia (11-13 November)

9	Harri Rovanperä/Risto Pietiläinen	2nd	KN04 WMD (Gp A)
10	Gianluigi Galli/Guido D'Amore	5th	KP54 GXY (Gp A)

Mitsubishi's WRC record in the Lancer Evolution era

This section shows how competitive the factory Mitsubishi team was compared to the other works outfits, and is also of interest when it comes to tracking the fortunes of the company's chief rivals mentioned elsewhere in the book.

1993 MAKERS
1. Toyota 157
2. Ford 145
3. Subaru 110
4. Mitsubishi 86
5. Lancia 75

1994 MAKERS
1. Toyota 151
2. Subaru 140
3. Ford 116
4. Mitsubishi 41

1995 MAKERS
1. Subaru 350
2. Mitsubishi 307
3. Ford 223
4. Toyota EXC

1996 MAKERS
1. Subaru 401
2. Mitsubishi 322
3. Ford 299

World champion: Tommi Mäkinen.

1997 MAKERS
1. Subaru 114
2. Ford 91
3. Mitsubishi 86

World champion: Tommi Mäkinen.

1998 MAKERS
1. Mitsubishi 91
2. Toyota 85
3. Subaru 65
4. Ford 53

World champion: Tommi Mäkinen.

1999 MAKERS
1. Toyota 109
2. Subaru 105
3. Mitsubishi 83
4. Ford 37
5. SEAT 23

World champion: Tommi Mäkinen.

(Continues overleaf)

2000 MAKERS
1. Peugeot 111
2. Ford 91
3. Subaru 88
4. Mitsubishi 43
5. SEAT 11

2001 MAKERS
1. Peugeot 106
2. Ford 86
3. Mitsubishi 69
4. Subaru 66
5. Skoda 17

2002 MAKERS
1. Peugeot 165
2. Ford 104
3. Subaru 67
4. Hyundai 10
5. Skoda 9
6. Mitsubishi 9

2003 MAKERS
1. Citroën 160
2. Peugeot 145
3. Subaru 109
4. Ford 93
5. Skoda 23

Mitsubishi had paused its WRC campaign.

2004 MAKERS
1. Citroën 194
2. Ford 143
3. Subaru 122
4. Peugeot 101

Mitsubishi finished the season unclassified.

2005 MAKERS
1. Citroën 188
2. Peugeot 135
3. Ford 104
4. Subaru 97
5. Mitsubishi 76

A 'LanEvo' swansong

Despite a Lancer Evolution estate, an electric version on test, and a Concept X prototype appearing well ahead of the 2006 season, rally fans (other than Paris-Dakar ones) were ultimately going to find themselves disappointed.

A press release issued on 14 December 2005 stated: "Mitsubishi Motors Corporation (MMC) has announced that it has decided to suspend participation in the FIA World Rally Championship series from 2006. The company hopes to return to the WRC series from 2008, after the completion of the three-year revitalization plan."

A clampdown on tobacco sponsorship, which had supported the rallying scene for decades, certainly didn't help, although DaimlerChrysler walking away from Mitsubishi a few weeks before the press statement was released may also have been a strong factor, as, with crazy budgets now being required to field a WRC team, finances were tight enough as it was, even before DaimlerChrysler withdrew.

Although the Mitsubishi's motorsports arm was now overseen from offices in Germany (MMSP GmbH), the company had always kept Andrew Cowan's old place in Rugby, England (later renamed MMSP Ltd) for the development side of the works rally cars, and it was MMSP Ltd that kept the Mitsubishi name represented in a semi-works effort over the next couple of years.

The 2006 season saw MMSP Ltd and Ralliart Italy support Gianluigi Galli's Lancer WRC05 for the opening round in Monte Carlo and the next one in Sweden. Galli sadly dropped out with gearbox problems on the Monte, but stayed the course in Sweden to claim fourth. Remarkably, Daniel Carlsson (partnered by Holmstrand Bosse) came third in a second WRC05 on his home event.

Jussi Välimäki was wheeled out for the Italian and Greek rounds

Jussi Välimäki on his way to seventh in the 2006 Rally Finland. *(Courtesy Antti Leppänen, Wikimedia Creative Commons)*

Gardemeister on the 2007 Monte Carlo Rally.
(Courtesy RobindesB, Wikimedia Creative Commons)

Finland, but Aava was the only one to finish in the top ten, claiming seventh. The Estonian driver was then sent to New Zealand, where he picked up eighth place.

The end of the line had been reached, though, and there were no Lancers entered during the rest of the season, and none competing in the 2008 World Rally Championship whatsoever. Soon after, John Easton completed a management buy-out to form a new enterprise called MML Sports Ltd.

The Lancers had definitely left their mark on the rally scene, though, claiming four titles for Tommi Mäkinen, and the manufacturer's crown in 1998. It also notched up 26 top class victories in the WRC, not to mention the numerous Group N wins. A true rally giant, if ever there was one ...

in the middle of the year, coming fifth in Sardinia, and ninth on the Acropolis. Finally, an all-out effort was made in Finland, with three Scandinavian drivers taking to the start. Välimäki was seventh, while Carlsson and Juho Hänninen failed to finish.

MMSP Ltd again fielded the WRC05 in 2007, running two cars for Toni Gardemeister and Xavier 'Xevi' Pons in the first three rounds. Gardemeister was seventh on the Monte, with Pons way down the field in 27th. Gardemeister moved up a spot in Sweden, but Pons went off the road to end his challenge. As the circus moved to Norway, the Mitsubishi men could only make up the numbers, with Pons 16th and Juho Hänninen 17th in a third car.

Gardemeister was excluded in Portugal, and Armindo Araujo went off the road in front of his home fans. At least the faith shown by the folks at MMSP Ltd was rewarded in Italy, with Gardemeister coming sixth, and Hänninen eighth in the second car. Later, Urmo Aava was signed up for the Acropolis, but 14th was the best he could muster. Hänninen and Aava were joined by Kristian Sohlberg and Kaj Kuistila in

The Mitsubishi service area pictured during the 2007 Rally Finland.
(Courtesty Yaamboo, Wikimedia Creative Commons)

More in the Rally Giants series:

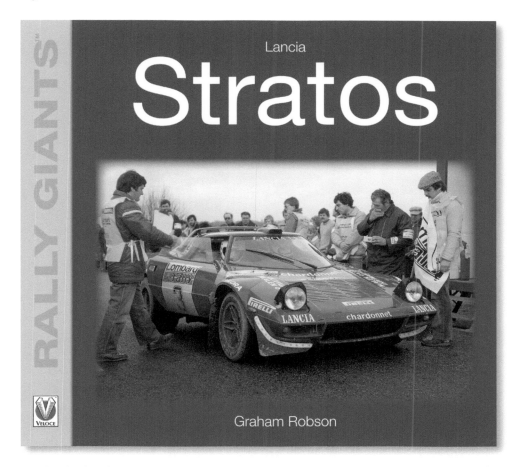

Describes the birth, development and rallying career of the Lancia Stratos, Europe's very first purpose-built rally car, in the mid/late 1970s. It provides a compact and authoritative history of where, when and how it became so important to the sport, as well as telling the story of the team.

ISBN: 978-1-787115-26-2
Paperback • 19.5x21cm • 128 pages • 131 colour and b&w pictures

For more information and price details, visit our website at www.veloce.co.uk
email: info@veloce.co.uk • Tel: +44(0)1305 260068

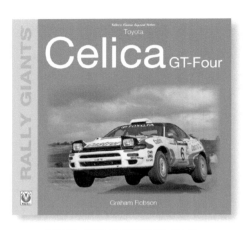

ISBN: 978-1-787113-31-2
Paperback • 19.5x21cm • 128
pages • 117 colour and b&w
pictures

ISBN: 978-1-787113-22-0
Paperback • 19.5x21cm • 128
pages • 96 colour and b&w
pictures

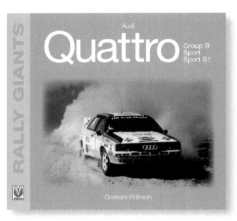

ISBN: 978-1-787111-10-3
Paperback • 19.5x21cm • 128
pages • 116 colour and b&w
pictures

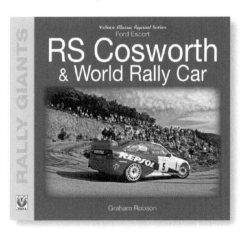

ISBN: 978-1-787111-71-4
Paperback • 19.5x21cm • 128
pages • 146 colour and b&w
pictures

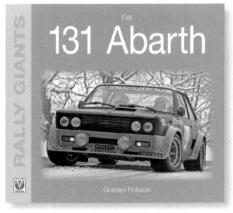

ISBN: 978-1-787111-11-0
Paperback • 19.5x21cm • 128
pages • 100 colour and b&w
pictures

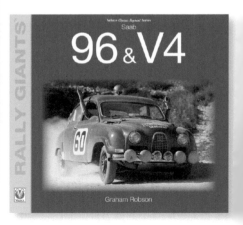

ISBN: 978-1-787113-32-9
Paperback • 19.5x21cm • 128
pages • 135 colour and b&w
pictures

From the running of the 1932 RAC rally, mainly a social event, to the present day, when Rally GB is a high-speed endurance World Championship rally, this is the very first all-embracing history of an important part of British motorsport history. Descriptions of every event, opinions, results and images are brought together for the very first time.

ISBN: 978-1-787117-36-5
Paperback • 22.5x22.5cm • 224 pages • 317 colour pictures

For more information and price details, visit our website at www.veloce.co.uk
email: info@veloce.co.uk • Tel: +44(0)1305 260068

INDEX

Aava, Urmo 122
Acropolis Rally (Greece) 6, 10, 11, 17, 19, 23-26, 37, 41, 48, 51, 55, 58, 63, 69, 71, 77, 83, 87, 90, 94, 96, 98, 103, 108, 112, 115, 117, 122
Aghini, Andrea 28, 30-34
Airikkala, Pentti 7, 11
Alim, Chandra 41
Amblard, Marc 49
Amigo Colon, Xavier 108, 109
Andersson, Ove 32
Anderson, Tord 33, 41
Arai, Toshihiro 65, 113
Araujo, Armindo 122
Argentina, Rally of 25, 26, 35, 37, 39, 41, 44, 48, 51, 55, 58, 63, 67, 69, 71, 73, 77, 87, 90, 94, 98, 103, 106, 108, 112, 117
Audi 68
Auriol, Didier 35, 38, 39, 41, 57, 63, 65, 68
Australia, Rally of 11, 19, 20, 28, 33-35, 37, 38, 41, 48, 50, 51, 54, 55, 57, 59, 64, 65, 69, 74, 77, 86, 88, 91, 97, 99, 104, 113, 115, 116, 118

Bäcklund, Kenneth 33, 35, 41, 48
Bargery, Andrew 99
Baskoro, Agung 41
Billstam, Claes 10
Bosse, Holmstrand 121
Burns, Richard 35-38, 41, 45, 47, 50, 51, 54-59, 62, 63, 65, 68, 70, 88, 89, 94

Campos, Miguel 70
Carlsson, Daniel 121, 122
Catalunya Rally (Spain) 30, 33, 34, 38, 41, 48, 51, 53, 55, 58, 63, 67, 68, 70, 72, 77, 87, 90, 93, 95, 98, 104, 109, 113, 118
Chandler, Morrie 6
China, Rally of 65, 69
Christie, Martin 25, 33, 34
Chrysler 13
Citroën 68, 92, 101, 112, 120
Climent, Luis 55
Cowan, Andrew 6, 32, 41, 121

Cyprus Rally 7, 72, 76, 77, 83, 87, 90, 93, 98, 103, 105, 108, 112, 117

D'Amore, Guido 108, 109, 117, 118
DaimlerChrysler 121
Dale, Justin 97, 99
Dauncey, Derek 96
Delecour, François 92-99
Dodge 13
Doig, David 6
Duez, Marc 62
Duncan, Ian 26
Dunkerton, Ross 11, 19, 20

Eagle 13
Easton, John 122
Ekstrom, Mattias 103
Enkei 65
Ericsson, Mikael 10
Eriksson, Kenneth 11, 17-22, 24-26, 28-35
Essen Show 102
Estrada, Roger 110

Farnocchia, Sauro 33, 34
Ferreyros, Ramón 89, 94
FFD Ricardo 49
FIA 12, 24, 33, 47, 57, 61, 80, 89, 100, 110, 113, 121
Ford 17, 26, 32, 48-50, 55, 63, 68, 72, 75, 87-89, 92, 96, 103, 112, 119, 120
Fornaris, Mario 100
Fujimoto, Yoshio 20, 57

Galli, Gianluigi 102-104, 106, 108-114, 116-118, 121
Gardemeister, Toni 88, 122
Giraudet, Denis 41
Gocentas, Fred 19
Grataloup, Daniel 92, 97-99
Grimaldi Forum 85
Grist, Nicky 19, 24
Grönholm, Marcus 62, 63, 68, 70, 74, 94

Hänninen, Juho 122
Hantunen, Timo 91
Harjanne, Seppo 25, 33, 34, 38, 41, 49-51
Harkki, Olli 25
Hartono, Bambang 41
Holderied, Isolde 24-26, 32-34
Hong Kong-Beijing Rally 33, 35, 40
Honkanen, Jakke 101
Hyundai 13, 92, 97, 120

Indonesia, Rally of 37, 41, 48, 51
Iritani, Kiyohito 48
Ishii, Shigehisa 13
Ivory Coast Rally 6-8, 10, 11

Julia, Oriol 104

Kankkunen, Juha 24, 37, 57, 65, 68, 88, 97
Kapanen, Arto 25, 98, 99, 101
Kasiman, Prihatin 41
Kataoka, Yoshihiro 36, 57
Kimata, Iwao 7, 100
Kobayashi, Kazuyoshi 13
Kondo, Akira 48
Kremer, Armin 101, 107
Kuistila, Kaj 122
Kulläng, Anders 6, 7
Kuukkala, Pentti 19, 25, 41
Kytolehto, Jarmo 25

Laine, Antero 7, 9
Lampi, Lasse 11, 41, 55, 79
Lancia 11, 17, 19, 20, 68, 119
Liatti, Piero 48
Lindauer, Bernard 49, 62, 81, 84, 96
Lindström, Kaj 88, 91, 108
Lloyd, Roland 31, 49
Loeb, Sebastien 112
Loix, Freddy 61-63, 65, 66, 68-77, 80, 82-84, 86-92
Lyon, Peter 104

Madeira, Rui 33, 34
Mäkinen, Tommi 11, 25, 26, 29, 31-41, 44-60, 62-74, 76, 77, 79,
    82-92, 94, 119, 122
Malaysia, Rally of 33
Mannisenmäki, Risto 58, 59, 68, 69, 76, 77, 88, 90, 91

Marlboro 100
Martini 19
Mazda 11
McNamee, Ronan 11
McRae, Alister 92-101
McRae, Colin 10, 20, 32, 33, 37, 48, 50, 55, 57, 63, 87-89, 92-94
McRae, Jimmy 10, 92
Menzi, Claudio 70
Mercedes-Benz 6
Mexico, Rally of 103, 104, 108, 111, 117
Michelin 109
Millbrook Proving Ground 82
Milles Pistes 8
Mitsubishi Oils 17, 26
MML Sports Ltd 122
Monte Carlo Rally 10-12, 17-19, 22, 25, 26, 29, 32, 33, 35, 44, 47,
    50, 55, 58, 60, 62, 68, 71-73, 76, 80, 82, 86, 90, 92, 93, 98,
    102-104, 108-111, 117, 121, 122

New Zealand, Rally of 6, 8, 10, 11, 19, 20, 23, 25, 26, 30, 32, 34,
    47, 48, 50, 57, 58, 63, 64, 69, 71, 75, 77, 85, 88, 90, 99, 101,
    103, 105, 108, 112, 117, 122
Nissan (Datsun) 11, 32
Nittel, Uwe 40, 41, 43, 48, 50, 51, 55, 71

Officer, David 8
Officer, Kate 8
Olympus Rally 10
Opel 6
Ordynski, Ed 34, 41, 50

Paasonen, Jani 72, 93, 95-101, 104, 106
Panizzi, Gilles 102-118
Panizzi, Hervé 108, 109, 117, 118
Papin, Claude 11
Paris-Dakar Rally 7, 8, 121
Park, Michael 112
Parmander, Staffan 19, 25, 33, 34
Peugeot 68, 75, 89, 93, 94, 96, 100, 103, 106, 112, 120
Pietlainen, Risto 117, 118
Plymouth 13
Pons, Xavier 104, 106, 122
Portugal, Rally of 11, 17-19, 25, 26, 32, 33, 46, 48, 51, 55, 58, 62,
    63, 68, 70, 77, 80, 83, 87, 90, 122
Pozzo, Gabriel 71, 87, 89
Prodrive 50

Proton 13, 26

Quandt, Sven 100, 103

RAC Rally (GB) 7-11, 17, 19-26, 33, 34, 49-51, 53-55, 57, 59, 66, 68, 69, 74, 76, 77, 88, 89, 91, 97, 99, 112, 118
Rådström, Thomas 55, 81, 87, 90
Ralliart 6-8, 11, 17, 26, 29, 31-33, 35, 37, 47-49, 54, 57, 62, 63, 70, 72, 74, 80-82, 88, 89, 92, 94, 96, 97, 100, 103, 121
Rally d'Italia Sardegna 104, 112, 113, 117, 122
Rally Japan 104, 112, 114, 118
Rally Norway 122
Rallye Deutschland 96, 99-101, 104, 107, 108, 112, 114, 118
Rautiainen, Timo 68
Recalde, Jorge 25, 26, 33, 34
Reid, Robert 41, 50, 51, 58, 59
Repo, Juha 41
Rothmans 32
Rovanperä, Harri 110-115, 117, 118

Safari Rally 6, 11, 17-19, 23, 25, 26, 32, 35, 37, 41, 45, 48, 50, 55, 56, 58, 62, 66, 68, 70, 73, 76, 80, 87, 89, 90, 94, 96, 98
Sainz, Carlos 20, 37, 48-50, 57, 62, 65, 68, 72, 87, 88, 94, 96
Salim, Eugene 7, 8
Salonen, Timo 11, 17
San Remo Rally 7, 20, 25, 26, 38, 39, 41, 50-52, 55, 57, 59, 65, 66, 69, 72, 75, 77, 80, 84, 85, 87, 88, 91, 96, 97, 99
Savignoni, Dominique 99
Schwarz, Armin 17-21, 23-26, 31
SEAT 119, 120
Senior, David 92, 98, 99, 101
Shinozuka, Kenjiro 6, 10, 11, 18-20, 23, 25, 26, 32, 35, 37, 41, 42
Short, Phil 31, 36
Singh, Joginder 6
Silva, Nuno 33, 34
Skoda 97, 120
Smeets, Sven 68, 69, 76, 77, 90, 91
Sohlberg, Kristian 100, 101, 103, 104, 106, 108, 122
Sola, Daniel 103, 104, 106, 107-109
Stacey, Mark 34, 41
Stenroos, Jyrki 41

Stohl, Manfred 48, 55, 57, 70-72, 74, 88, 104, 106
Subaru 11, 14, 20, 26, 32, 33, 36-38, 48, 50, 57, 62, 65, 68, 70, 75, 89, 92, 97, 101, 102, 104, 112, 113, 119, 120
Swedish Rally 17, 29, 32, 33, 35, 37, 41, 48, 50, 55, 56, 58, 62, 68, 70, 72, 76, 81, 87, 90, 93, 94, 98, 100, 101, 103, 108, 111, 112, 117, 121, 122

Tanaka, Yasuo 110
Tauziac, Patrick 8, 11
Thailand, Rally of 33
Thiry, Bruno 97
Thorner, Tina 25, 26, 32, 33, 34, 41, 50, 51, 90
Thul, Peter 19
Tokito, Saburo 42
Tokyo Show 13, 52, 70, 102
Torii, Isao 110
Tour de Corse 25, 26, 28, 32, 34, 43, 48, 51, 55, 58, 63, 69, 72, 88, 91-93, 98, 104, 112, 114, 118
Toyota 11, 17, 19, 20, 24, 26, 31-33, 55, 57, 62, 63, 65, 68, 119
Trelles, Gustavo 40, 48, 50, 55, 57, 63, 65, 68, 72, 74, 77, 87, 88
TTE 17, 32, 55, 68
Turkey, Rally of 103, 106, 108, 112, 117

Valimaki, Jussi 113, 121, 122
Vatanen, Ari 10, 11, 33
Virjula, Antti 25

Wicha, Klaus 24, 25

Yace, Martial 8
Yokoyama, Taizo 48

Zanirolli, Patrick 8

1000 Lakes Rally (Finland) 7, 9, 11, 17, 19, 20, 25, 26, 32, 35, 37, 41, 45, 48, 51, 55, 57, 58, 65, 69, 72, 74, 77, 84, 88, 90, 94, 97, 98, 104, 108, 112, 117, 121, 122

The Mitsubishi company and its products are mentioned throughout this book.